T0233299

Lecture Notes in Computer Science 8991

Commenced Publication in 1973
Founding and Former Series Editors:
Gerhard Goos, Juris Hartmanis, and Jan van Leeuwen

More information about this series at http://www.springer.com/series/7409

Tilmann Rabl · Kai Sachs
Meikel Poess · Chaitanya Baru
Hans-Arno Jacobson (Eds.)

Big Data Benchmarking

5th International Workshop, WBDB 2014
Potsdam, Germany, August 5–6, 2014
Revised Selected Papers

 Springer

Editors

Tilmann Rabl
University of Toronto
Toronto, ON
Canada

Kai Sachs
SAP SE
Köln
Germany

Meikel Poess
Server Technologies
Oracle Corporation
Redwood Shores, CA
USA

Chaitanya Baru
University of California at San Diego
La Jolla, CA
USA

Hans-Arno Jacobson
Middleware Systems Research Group
Toronto, ON
Canada

ISSN 0302-9743   ISSN 1611-3349   (electronic)
Lecture Notes in Computer Science
ISBN 978-3-319-20232-7   ISBN 978-3-319-20233-4   (eBook)
DOI 10.1007/978-3-319-20233-4

Library of Congress Control Number: 2015941491

LNCS Sublibrary: SL3 – Information Systems and Applications, incl. Internet/Web, and HCI

Springer Cham Heidelberg New York Dordrecht London

Printed on acid-free paper

Springer International Publishing AG Switzerland is part of Springer Science+Business Media
(www.springer.com)

Preface

Formed in 2012, the Big Data Benchmarking Community (BDBC) represents a major step in facilitating the development of benchmarks for objective comparisons of hardware and software systems dealing with emerging big data applications. Led by Chaitanya Baru, Tilmann Rabl, Meikel Poess, Milind Bhandarkar, and Nambiar Raghunath, the BDBC has successfully conducted five international Workshops on Big Data Benchmarking (WBDB). One strength of the WBDB is that it brings together practitioners and researchers, which leads to a balance between industrial and academic contributions. It provides the right environment in which to discuss the challenges and potential approaches to benchmark big data systems. The results of the WBDB have a high impact on the ongoing research in the domain and WBDB itself is established as the leading event focusing on big data benchmarking. Previous editions in 2012 and 2013 were organized in the USA, India, and China. In August 2014, the fifth WBDB finally came to Europe.

WBDB 2014 took place in Potsdam hosted by the Hasso Plattner Institute. It was only possible thanks to the generous support of SAP SE and HP/Suse as gold sponsors and Intel, Mellanox, and Pivotal as silver sponsors. The number of high-quality submissions allowed us to define an interesting program covering many different aspects of big data benchmarking. Each paper was reviewed by at least three reviewers. In addition to the technical program, the local arrangements team under Matthias Uacker did a great job and made it an unforgettable event. The program of the workshop was enriched with two exciting keynote talks:

- "An Approach to Benchmarking Industrial Big Data Application" by Umseh Dayal
- "A TU Delft Perspective on Benchmarking Big Data in the Data Center" by Alexander Iosup.

Beside all the exciting talks, we had many inspiring discussions on the different presentations, e.g., on how to build standard benchmarks that can be used by industry as well as academia. In this book, we collected the revised version of the submissions and grouped them into the following three sections: "Benchmark Specifications and Proposals"—submissions focusing on concrete benchmarks and/or proposals on how to implement such benchmarks, "Hadoop and MarReduce"—papers discussing Hadoop and MapReduce in the different context such as virtualization and cloud, and "In-Memory, Data Generation Graphs."

Overall, 2014 was an exciting year for the BDBC. An important step in the direction of an industry standard benchmarks for big data benchmarking was the decision of the BDBC to join the Standard Performance Evaluation Corporation (SPEC). The BDBC formed a new working group within the SPEC Research Group (RG). The SPEC RG is a group within SPEC established to serve as a platform for collaborative research efforts in the area of quantitative system evaluation and analysis, fostering the interaction between industry and academia in the field. The collaboration between the

BDBC and SPEC was announced at WBDB 2014. Since 2015, the BDBC community has been represented in the SPEC RG Steering Committee by Tilmann Rabl, who is also heading the Big Data Benchmarking working group together with Chaitanya Baru. The working group is currently working on a proposal for an industry standard benchmark based on the BigBench proposal. Further details about BigBench are discussed in this book.

We thank the sponsors, members of the Program Committee, authors, local arrangement team, and participants for their contributions to the workshop.

March 2015   Chaitanya Baru
                                                             Tilmann Rabl
                                                               Kai Sachs

Organization

Organizing Committee

General Chairs

Chaitanya Baru   UC San Diego, USA
Tilmann Rabl   University of Toronto, Canada
Kai Sachs   SAP SE, Germany

Local Arrangements

Matthias Uacker   Hasso Plattner Institute, Germany

Publicity Chair

Henning Schmitz   SAP Innovations Center, Germany

Publication Chair

Meikel Poess   Oracle, USA

Program Committee

Milind Bhandarkar   Pivotal, USA
Anja Bog   SAP Labs, USA
Dhruba Borthakur   Facebook, USA
Joos-Hendrik Böse   Amazon, USA
Tobias Bürger   Payback, Germany
Tyson Condi   UCLA, USA
Kshitij Doshi   Intel, USA
Pedro Furtado   University of Coimbra, Portugal
Bhaskar Gowda   Intel, USA
Goetz Graefe   HP Labs, USA
Martin Grund   University of Fribourg, Switzerland
Alfons Kemper   TU München, Germany
Donald Kossmann   ETH Zurich, Switzerland
Tim Kraska   Brown University, USA
Wolfgang Lehner   TU Dresden, Germany
Christof Leng   UC Berkeley, USA
Stefan Manegold   CWI, The Netherlands
Raghunath Nambiar   Cisco, USA
Manoj Nambiar   Tata Consulting Services, India
Glenn Paulley   Conestoga College, Canada

Scott Pearson	Cray, USA
Andreas Polze	Hasso Plattner Institute, Germany
Alexander Reinefeld	Zuse Institute Berlin/HU Berlin, Germany
Berni Schiefer	IBM Labs Toronto, Canada
Saptak Sen	Hortonworks, USA
Florian Stegmaier	University of Passau, Germany
Till Westmann	Oracle Labs, USA
Jianfeng Zhan	Chinese Academy of Sciences, China

Sponsors

Platinum:	SAP SE
	HP/Suse
Gold:	Intel
	Mellanox
	Pivotal
	US National Science Foundation
In-Kind:	Hasso Plattner Institute
	San Diego Supercomputer Center

Contents

Benchmark Specifications and Proposals

Towards a Complete BigBench Implementation

Tilmann Rabl[1,2]($\boxtimes$), Michael Frank[2], Manuel Danisch[2], Bhaskar Gowda[3],
and Hans-Arno Jacobsen[1]

[1] Middleware Systems Research Group, University of Toronto, Toronto, ON, Canada
`tilmann.rabl@utoronto.ca`, `jacobsen@eecg.toronto.edu`
[2] Bankmark UG, Passau, Germany
{`michael.frank,manuel.danisch`}`@bankmark.de`
[3] Intel Corporation, Santa Clara, CA, USA
`bhaskar.d.gowda@intel.com`

Abstract. BigBench was the first proposal for an end-to-end big data analytics benchmark. It features a set of 30 realistic queries based on real big data use cases. It was fully specified and completely implemented on the Hadoop stack. In this paper, we present updates on our development of a complete implementation on the Hadoop ecosystem. We will focus on the changes that we have made to data set, scaling, refresh process, and metric.

1 Introduction

Modern storage technology enables storing more and more detailed information. Retailers are able to record all user interaction to improve shopping experience and marketing. This is enabled by new big data management systems. These are new storage systems that make it possible to manage and process ever larger amounts of data. To date a plethora of different systems is available featuring new technologies and query languages. This makes it hard for customers to compare features and performance of different systems. To this end, BigBench, the first proposal for an end-to-end big data analytics benchmark [1] was developed. BigBench was designed to cover essential functional and business aspects of big data use cases.

In this paper, we present updates on our alternative implementation of the BigBench workload for the Hadoop eco-system. We reimplemented all 30 queries and the data generator. We adapted metric, scaling, data set, and refresh process the fit the needs of a big data benchmark.

The rest of the paper is organized as follows. In Sect. 2, we present a brief overview of the BigBench benchmark. Section 3 introduces the new scaling model for BigBench. We give details on refresh process in Sect. 4. Section 5 presents our proposal for a new big data metric. Section 6 gives an overview of related work. We conclude with future work in Sect. 7.

T. Rabl et al. (Eds.): WBDB 2014, LNCS 8991, pp. 3–11, 2015.
DOI: 10.1007/978-3-319-20233-4_1

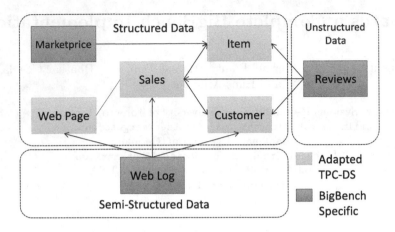

Fig. 1. BigBench schema

2 BigBench Overview

BigBench is based on the Transaction Processing Performance Council's (TPC) new decision support benchmark TPC-DS [2,3]. TPC-DS models a retail business with store, online, and catalog market channels. For BigBench the catalog channel was dropped and additional big data related information sources were added. The reason for this was that catalog sales are becoming ever less important in current retail businesses and more and more additional information sources are included in big data analytics. A high-level overview of the extended data model can be seen in Fig. 1. The picture shows the sales channel, which in the full schema consists of web sales, web returns, store sales, and store returns. The dimension marketprice was added to store competitors' prices. The Web Log portion represents a click stream that records user behavior on the online presence. The click stream is stored in comma separated value form but resembles a web server log. The product reviews contain full text that is used for natural language processing. The complete schema is described in [4]. In this new version of the data generator, the complete data set was implemented using the parallel data generation framework [5].

BigBench features 30 complex queries, 10 of which are taken from TPC-DS. The queries are covering the major areas of big data analytics [6]. As a result, they cannot be expressed by pure SQL queries. The full list of queries in Teradata Aster SQL-MR syntax can be found in [4]. In the current version, all queries were implemented using Hadoop, Hive, Mahout, OpenNLP [7].

3 BigBench Scaling

As the original BigBench implementation was partially based on the TPC-DS data generator DSDGen, it was bound to the same restrictions in terms of scaling. DS achieves perfect data size scalability, i.e., data is exactly scaling as specified by the scale factor scale. This is done by restricting the scale factor to certain numbers (100, 300, 1000, 3000, 10000, 30000, 100000) and fine tuning the table sizes to the correct data size for each scale factor. This limits the maximum data size to 100TB. Furthermore, there are no intermediate data sizes, which makes it hard to test load boundaries of a system or testing the optimal load factor. Therefore, we have redesigned the scaling to allow for a continuous scaling.

Simply using a linear scale factor for all table sizes is not an option for BigBench, since it is supposed to scale across a very wide range of scale factors. Our initial target is to have reasonable table sizes for all tables for the range of 1 GB to 1 PB, allowing for small and fast experimental tests up to large scale out experiments. This means that the number of entries of a linearly scaled table will be increased by a factor of 10^6 for the 1 PB data set. This is not realistic for many of the tables, for example, it is unlikely that a company has more than 50 K stores worldwide. For comparison, Walmart, the world's largest retailer has 11 K stores[1], McDonald's has 33 K restaurants worldwide[2], Starbucks has 17 K stores worldwide[3]. Consequently, several tables in the schema cannot be scaled linearly. TPC-DS uses linear, logarithmic, and square root order of growth for different tables but the individual numbers are hand adjusted. We use the same approach but uniformly adjust the linear scaling factors to make up for the non-linear growth of other tables. An overview of all tables and their growth factors can be seen in Table 1.

To adjust the sublinear growth of some of the tables, we increase the linear scaling factor by the percentage that is missing to get a linear scaling of the data size. This can be computed by measuring the relative data size of each different class of growth factor for scale factor 1. For example, the linearly scaled tables are contributing roughly 50 % of the data to the 1 GB data set. The rest comes from static and sublinearly scaling tables. The strongest growing table of the latter is inventory, which contains a record for every item in every warehouse every day. To make up for the missing data size we compute the difference in growth between the linear factor and the sublinear factors in comparison to the base size and increase the linear factor by this factor:

$$LF = SF + (SF - (\log_5(SF) * \sqrt{SF})) = 2SF - \log_5(SF) * \sqrt{SF} \qquad (1)$$

where LF is the linear factor and SF is the scale factor. For large scale factors the linear factor converges towards $2SF$ and thus the linearly scaling tables make up almost the complete data size as expected.

4   Refresh

The refresh process initially specified in BigBench is an exact copy of the TPC-DS refresh [1]. TPC-DS mandates the refresh to happen during a throughput

---

[1] Walmart Interactive Map - http://corporate.walmart.com/our-story/our-business/locations/#/.

[2] McDonald's FAQs http://www.mcdonalds.ca/ca/en/contact_us/faq.html.

[3] Starbucks Company Profile - http://www.starbucks.com/about-us/company-information.

Table 1. Overview of table growth factors

Table name	# Rows SF 1	Bytes/Row	Scaling
date	109573	141	static
time	86400	75	static
ship_mode	20	60	static
Household_demographics	7200	22	static
customer_demographics	1920800	40	static
customer	100000	138	square root
customer_address	50000	107	square root
store	12	261	square root
warehouse	5	107	logarithmic
promotion	300	132	logarithmic
web_page	60	134	logarithmic
item	18000	308	square root
item_marketprice	90000	43	square root
inventory	23490000	19	square root * logarithmic
store_sales	810000	143	linear
store_returns	40500	125	linear
web_sales	810000	207	linear
web_returns	40500	154	linear
web_clickstreams	6930000	27	linear
product_reviews	98100	670	linear

test. Given S streams of query sequences, each simulating a user, there are $S/2$ refresh operations each scheduled after 198 queries are completed in total in the streams (each stream runs 99 queries for a total of $99 * S$ queries in all streams). This was added to TPC-DS later to ensure that systems are able to deal with trickle updates and do not "over-optimize" the storage layout. Many big data systems, hovever, process data in a batch oriented fashion. In this model, data is either completely loaded fresh after new data has to be processed or updates are processed in bulk as is also common in many data warehouses. In current Hive, for instance, any update of the data means overwriting complete files, since files in HDFS are immutable. Therefore, we have changed the refresh model to a single bulk update in between two throughput runs. This ensures that refresh has to be handled by the system (even if it means a complete reload of the data) but at the same time the refresh will not enforce stopping a throughput run, which would be the case using TPC-DS' update mechanism. By default, new data in the size of 1 % of the original data set is inserted in the refresh phase. The size of the refresh data for each table is determined by the scaling model described in Sect. 3.

5 Metric

We propose a new metric for BigBench to take the batch-oriented processing of many big data systems into account and to include all parts of the benchmark. As BigBench aims at becoming an industry standard benchmark, we require a combined metric that returns a single (abstract) value that can be used to compare the entire end-to-end performance of big data analytics systems. The initially proposed metric for BigBench was loosely based on the TPC-DS metric. It consisted of four measurements:

T_L: Execution time of the loading process;
T_D: Execution time of the declarative queries;
T_P: Execution time of the procedural queries;
T_M: Execution time of queries with procedural and declarative aspects.

The complete metric was specified as the geometric mean of these four measurements:

$$\text{Metric} = \sqrt[4]{T_L * T_D * T_P * T_M} \tag{2}$$

The intuition behind the metric was that some systems are optimized for declarative queries while others are optimized for procedural queries and the classification can give an indication on the type of system that is benchmarked. There are a couple of problems with this metric, most notably the classification of queries, which is debatable. Because the concrete implementation of queries is not enforced the classification does not uniformly apply. For example, in Hive all queries will be transformed to MapReduce jobs and thus finally be run in a procedural way. Furthermore, the metric does not consider the refresh process or parallel execution of queries. To address these issues, we have revisited the TPC-H [8] and TPC-DS metrics [3]. To explain the reasoning behind the new metric, we will give a short overview of these two metrics as specified in the current versions. Both, TPC-H and TPC-DS specify two performance tests, a power test and a throughput test. The resulting numbers are meant to be comparable for a certain database size. The power test represents a single user run and shows the ability of a system to run a single stream of queries. The throughput test represents a multi user scenario where many queries are executed in parallel.

In TPC-H the power test is specified as a serial run of all 22 queries and one refresh process before and one after the query run. The power metric is defined as follows:

$$\text{Power\_Metric@Size} = \frac{3600 * SF}{\sqrt[24]{T_{R_1} * \prod_{i=1}^{22} T_{Q_i} * T_{R_2}}} \tag{3}$$

where T_x is the processing time of a query or refresh function in seconds. The total metric is the geometric mean of the number of queries and refreshes that can be processed per hour multiplied by the scale factor. The reasoning behind multiplying with the scale factor is probably a higher comparability between different scale factors, since the data size in GB equals the scale factor. The throughput test is specified as a run of S serial streams of each a permutation of the 22 queries and a separate stream of $2S$ refresh functions. The scheduling of the refresh stream is not specified. The throughput metric is specified as follows:

$$\text{Throughput\_Metric@Size} = \frac{S * 22 * 3600 * SF}{T_S} \qquad (4)$$

where S is the number of streams and T_S is the total execution time of the throughput test. The metric is similar to the power metric but it computes the arithmetic mean of the processing times. It is unclear why the refresh functions are not considered in the arithmetic mean. Again the resulting average number of queries per hour is multiplied with the scale factor.

TPC-H does not consider the loading time as part of the metric. The final metric is specified as the geometric mean of the power metric and the throughput metric:

$$\text{Queries\_per\_hour@Size} = \sqrt{\text{Power\_Metric@Size} * \text{Throughput\_Metric@Size}}$$

$$(5)$$

TPC-DS has the same possible variables, the number of streams in the throughput test S and the scale factor SF. The TPC-DS metric consists of 3 parts:

T_{LD}: The load factor;
T_{PT}: The power test factor;
T_{TT}: The throughput factor.

The load factor is computed by measuring the time it takes to load the data T_{load} multiplied with the number of stream S in the throughput test and a factor 0.01: $T_{LD} = T_{\text{load}} * S * 0.01$.

The power test is a serial run of all 99 queries of TPC-DS. There is no data refresh in the power test in TPC-DS. The power test factor is the total execution time of the power test multiplied by the number of streams: $T_{PT} = T_{\text{Power}} * S$.

The throughput test consists of two independent, consecutive runs of S parallel query streams interspersed with $S/2$ refresh functions. Each of the streams executes all 99 queries in a permutation specified by TPC-DS. The reported time is the total execution time of both runs. All times are reported in hours.

The complete DS metric is specified as follows:

$$\text{QphDS@SF} = \left\lfloor \frac{SF * 3 * S * 99}{T_{LD} + T_{PT} + T_{TT}} \right\rfloor \qquad (6)$$

Note that this metric is different than the initially presented metric for TPC-DS as specified in [2]. The initially specified metric did not contain the power test and specified a separate data maintenance phase between two query runs:

$$\text{QphDS@SF}_{\text{old}} = \left\lfloor \frac{SF * S * 198}{T_{LD} + T_{TT_1} + T_{DM} + T_{TT_2}} \right\rfloor \qquad (7)$$

where T_{TT_1} and T_{TT_2} are the run times of the 2 throughput test runs and T_{DM} is the run time of the data maintenance. This metric was designed to compute the

average query throughput per hour including maintenance tasks. Based on these
two metrics, we have designed a new metric that takes the application scenario
of BigBench into account. The metric includes the following components:

T_L: Execution time of the loading process;
T_P: Execution time of the power test;
T_{TT_1}: Execution time of the first throughput test;
T_{DM}: Execution time of the data maintenance task.
T_{TT_2}: Execution time of the second throughput test;

All times are measured in seconds in at least 0.1 precision. In TPC-H loading is
not measured at all, which is only valid if loading is a very rare activity in the life-
cycle of a database. In TPC-DS a 1 % fraction of the load time is incorporated
in the metric, which is then multiplied by the number of streams to increase
the impact of the effort of loading for increasing number of streams. This still
implies loading is a rare operation in the targeted application. Since we believe
that loading is an essential part of the workload in big data applications, we
keep the full loading time in the metric. The finally reported metric is similar to
the initial metric of TPC-DS:

$$\text{BBQpH} = \frac{30 * 3 * S * 3600}{S * T_L + S * T_P + T_{TT_1} + S * T_{DM} + T_{TT_2}} \tag{8}$$

The metric reports the mean queries per hour including the loading time and
data maintenance time. The times are multiplied with the number of streams in
order to keep the impact of the power test, the loading, and the data maintenance
stable. The total number of queries run is $30 * (2 * S + 1)$, but since the run time
of the power test is multiplied by S, the 30 queries in that run are counted S
times. Therefore the numerator of the metric is $30 * 3 * S$ times the number of
seconds in an hour.

6   Related Work

Today, TPC benchmarks are commonly used for benchmarking big data systems.
For big data analytics, TPC-H and TPC-DS are obvious choices and TPC-H has
been implemented in Hadoop, Pig[4], and Hive[5] [9,10]. Subsets of TPC-DS queries
have been implemented in Impala[6], Hive, Hawq, Shark, and others. TPC-H and
TPC-DS are pure SQL benchmarks and do not cover all the different aspects
that MapReduce systems are typically used for.

Several proposals try to modify TPC-DS similar to BigBench to cover typical
big data use cases. Zhao et al. propose Big DS, which extends the TPC-DS model
with social marketing and advertisement [11]. To resemble the ETL part of a
big data workload, Yi and Dai have modified TPC-DS as part of the HiBench

---

[4] https://issues.apache.org/jira/browse/PIG-2397.
[5] https://issues.apache.org/jira/browse/HIVE-600.
[6] http://blog.cloudera.com/blog/2014/01/impala-performance-dbms-class-speed/.

suite [12,13]. The authors use the TPC-DS model to generate web logs similar to BigBench and use custom update functions to simulate an ETL process. Like BigDS this is orthogonal to BigBench and can be included in future versions of the benchmark. There have been several other proposals like the Berkeley Big Data Benchmark[7] and the benchmark presented by Pavlo et al. [14]. Another example is BigDataBench, which is a suite similar to HiBench and mainly targeted at hardware benchmarking [15]. Although interesting and useful, both benchmarks do not reflect an end-to-end scenario and thus have another focus than BigBench.

7   Conclusion

BigBench is the only fully specified end-to-end benchmark for big data analytics currently available. In this paper, we present updates on the data model, refresh, metric, and scaling. The queries and the data set can be downloaded from GitHub[8].

References

1. Ghazal, A., Rabl, T., Hu, M., Raab, F., Poess, M., Crolotte, A., Jacobsen., H.A.: BigBench: towards an industry standard benchmark for big data analytics. In: SIGMOD (2013)
2. Pöss, M., Nambiar, R.O., Walrath, D.: Why you should run TPC-DS: a workload analysis. In: VLDB, pp. 1138–1149 (2007)
3. Transaction Processing Performance Council: TPC Benchmark H - Standard Specification, Version 2.17.0 (2012)
4. Rabl, T., Ghazal, A., Hu, M., Crolotte, A., Raab, F., Poess, M., Jacobsen, H.-A.: BigBench specification V0.1. In: Rabl, T., Poess, M., Baru, C., Jacobsen, H.-A. (eds.) WBDB 2012. LNCS, vol. 8163, pp. 164–201. Springer, Heidelberg (2014)
5. Rabl, T., Frank, M., Sergieh, H.M., Kosch, H.: A data generator for cloud-scale benchmarking. In: Nambiar, R., Poess, M. (eds.) TPCTC 2010. LNCS, vol. 6417, pp. 41–56. Springer, Heidelberg (2011)
6. Manyika, J., Chui, M., Brown, B., Bughin, J., Dobbs, R., Roxburgh, C., Byers, A.H.: Big data: the next frontier for innovation, competition, and productivity. Technical report, McKinsey Global Institute (2011). http://www.mckinsey.com/insights/mgi/research/technology_and_innovation/big_data_the_next_frontier_for_innovation
7. Chowdhury, B., Rabl, T., Saadatpanah, P., Du, J., Jacobsen, H.A.: A BigBench implementation in the Hadoop ecosystem. In: Rabl, T., Raghunath, N., Poess, M., Bhandarkar, M., Jacobsen, H.-A., Baru, C. (eds.) ABDB 2013. LNCS, vol. 8585, pp. 3–18. Springer, Heidelberg (2014)
8. Transaction Processing Performance Council: TPC Benchmark DS - Standard Specification, Version 1.1.0 (2013)
9. Moussa, R.: TPC-H benchmark analytics scenarios and performances on Hadoop data clouds. In: Benlamri, R. (ed.) NDT 2012, Part I. CCIS, vol. 293, pp. 220–234. Springer, Heidelberg (2012)

---

7 https://amplab.cs.berkeley.edu/benchmark/.
8 https://github.com/intel-hadoop/Big-Bench.

10. Kim, K., Jeon, K., Han, H., Kim, S.G., Jung, H., Yeom, H.: MRBench: a benchmark for MapReduce framework. In: 14th IEEE International Conference on Parallel and Distributed Systems, ICPADS 2008, pp. 11–18, December 2008
11. Zhao, J.M., Wang, W., Liu, X.: Big Data Benchmark - Big DS. In: Proceedings of the Third and Fourth Workshop on Big Data Benchmarking 2013. (2014) (in print)
12. Huang, S., Huang, J., Dai, J., Xie, T., Huang, B.: The HiBench benchmark suite: characterization of the MapReduce-based data analysis. In: ICDEW (2010)
13. Yi, L., Dai, J.: Experience from Hadoop Benchmarking with HiBench: from Micro-Benchmarks toward End-to-End Pipelines. In: Proceedings of the Third and Fourth Workshop on Big Data Benchmarking 2013. (2014) (in print)
14. Pavlo, A., Paulson, E., Rasin, A., Abadi, D.J., DeWitt, D.J., Madden, S., Stonebraker, M.: A comparison of approaches to large-scale data analysis. In: Proceedings of the 35th SIGMOD international conference on Management of data SIGMOD 2009, pp. 165–178 (2009)
15. Wang, L., Zhan, J., Luo, C., Zhu, Y., Yang, Q., He, Y., Gao, W., Jia, Z., Shi, Y., Zhang, S., Zhen, C., Lu, G., Zhan, K., Li, X., Qiu, B.: BigDataBench: a big data benchmark suite from internet services. In: Proceedings of the 20th IEEE International Symposium On High Performance Computer Architecture. HPCA (2014)

BW-EML SAP Standard Application Benchmark

Heiko Gerwens and Tobias Kutning[(✉)]

SAP SE, Walldorf, Germany
tobas.kutning@sap.com

Abstract. The focus of this presentation is on the latest addition to the BW SAP Standard Application Benchmarks, the SAP Business Warehouse Enhanced Mixed Load (BW-EML) benchmark. The benchmark was developed as a modern successor to the previous SAP Business Warehouse benchmarks. With near real-time and ad hoc reporting capabilities on big data volumes the BW-EML benchmarks matches the demands of modern business warehouse customers. The development of the benchmark faced the challenge of two contradicting goals. On the one hand the reproducibility of benchmark results is a key requirement. On the other hand the variability in the query workload was necessary to reflect the requirements for ad hoc reporting. The presentation will give an insight to how these conflicting goals could be reached with the BW-EML benchmark.

1 SAP Standard Application Benchmarks

SAP SE is the world´s leading provider of business software[1]. It delivers a comprehensive range of software products and services to its customers: Companies from all types of industries, ranging from small businesses to large, multinational enterprises engaged in global markets.

For more than 20 years now SAP and its hardware and technology partners have developed and run benchmarks to test the performance and scalability of both hardware and the SAP solutions running on the hardware.

The first benchmark was certified in 1995. Since then the SAP Standard Application Benchmarks have become some of the most important benchmarks in the industry. Especially the SAP SD standard application benchmark[2] can be named as an example.

The goal of the SAP Standard Application Benchmarks is to represent SAP business applications as realistic as possible. A close to real life workload is one of the key elements of SAP benchmarks.

The performance of SAP components and business scenarios is assessed by the benchmarks and at the same time input for the sizing procedures is generated. Performance in SAP Standard Application Benchmarks is determined by throughput numbers and system response times. The throughput numbers are defined in business

---

[1] http://www.sap.com/about.html, http://www.softwaretop100.org/enterprise-software-top-10.
[2] http://global1.sap.com/campaigns/benchmark/appbm_sd.epx.

© Springer International Publishing Switzerland 2015
T. Rabl et al. (Eds.): WBDB 2014, LNCS 8991, pp. 12–23, 2015.
DOI: 10.1007/978-3-319-20233-4_2

application terms. For the SAP SD benchmark this would be for example fully processed order line items per hour. This business application throughput is then mapped onto the hardware resource consumption for example CPU and memory. The unit for the measurement of CPU power is SAP Application Performance Standard (SAPS)[3]. SAPS is a hardware independent measurement unit for the processing power of any system configuration. 100 SAPS is defined as 2,000 fully processed order lines items per hour which equals 2,000 postings or 2,400 SAP transactions.

By ensuring that the benchmark results are free of artifacts, the availability of the tested hardware and software combinations for productive use by customers SAP and its partners ensure that the benchmarks are highly business relevant for customers. The benchmarking methods are monitored and approved by the SAP Benchmark Council, which consists of SAP and its hard- and software partners.

Customers, SAP partners and also SAP profit from the SAP Standard Application Benchmarks. For customers the benchmarks can serve as a proof of concept illustrating the scalability and manageability of large SAP installations from both hard- and software point of view. Customers are also able to compare different hard- and software combinations and releases using the benchmark data.

Partners of SAP are enabled to optimize their technology for SAP applications with the help of the SAP Standard Application Benchmarks. The benchmarks are also used to prove the scalability of hardware components and of course the marketing departments do use the benchmark results to support sales.

For SAP the main benefits of the SAP Standard Application benchmarks are quality assurance, analysis of system configurations and parameter settings and the verification of sizing recommendations.[4]

2   Benchmark Simulation with the SAP Benchmark Tools

The SAP benchmark toolkit consists of collection of Perl scripts and C programs and is available for Windows, Linux and UNIX. In addition there are pre-configured SAP system environments available containing all necessary business data to execute the different benchmarks.

In the area of scripting the benchmark tools do allow the recording and playback of DIAG (the protocol used by the SAP frontend software SAP GUI) and HTTP protocol user interactions. The benchmark scripts can simulate a configurable number of system users executing a pre-defined series of transactions in the SAP system. The tools also take care of collecting all the necessary monitoring data from the collection tools built into the SAP System and also directly from the operating system. To check that the benchmark runs were free of errors the benchmark tools also collect the data returned for each system interaction of every user.

The Multi-User Simulation in SAP Standard Application Benchmarks can be divided in 3 sections Fig. 1 – the ramp up phase, the high load phase and the ramp-down

---

[3] http://global1.sap.com/campaigns/benchmark/measuring.epx.

[4] http://global1.sap.com/campaigns/benchmark/bob_overview.epx.

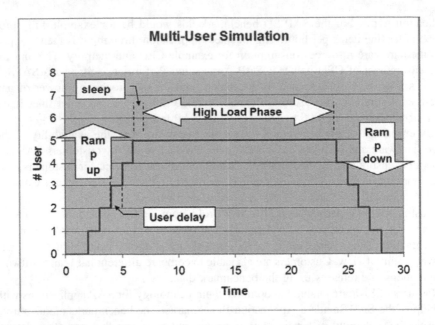

Fig. 1. Multi-user simulation in a SAP standard application benchmark

phase. In the ramp up phase all the configured users are logging into the system and start loops of their preconfigured steps of transactions. The speed with which users are logging onto the system can be adjusted using a user logon delay parameter. After all users are logged into the system the next phase called the high load phase starts. This is the phase where all configured users are logged into the system and perform the workload. For the SD benchmark the high load phase needs to be at least 15 min and for the BW-EML benchmark at least 1 h.[5]

3 SAP Business Warehouse Architecture

In order to understand the SAP terms used in the following discussion of the BW-EML benchmark let us take a look at some of the most important parts of the SAP BW architecture (Fig. 2).

3.1 InfoProvider

An InfoProvider is an object for which queries can be created or executed. InfoProviders are the objects or views that are relevant for reporting. The following types of InfoProviders[6] used in the BW-EML benchmark.

---

[5] http://global1.sap.com/campaigns/benchmark/bob_run.epx.

[6] http://help.sap.com/saphelp_sem60/helpdata/en/4d/c3cd3a9ac2cc6ce10000000a114084/content. htm?frameset=/en/8d/2b4e3cb7f4d83ee10000000a114084/frameset.htm&current_toc=/en/a3/ fe1140d72dc442e10000000a1550b0/plain.htm&node_id=85&show_children=false.

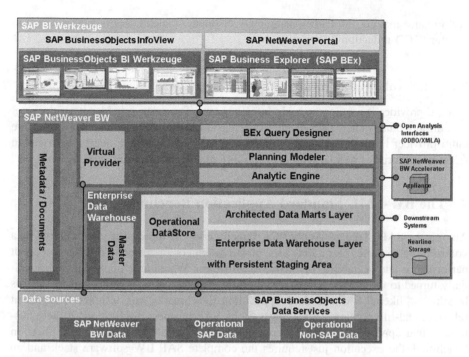

Fig. 2. SAP BW architecture overview (source: http://help.sap.com/erp2005_ehp_07/helpdata/
en/47/5fa4468d0268b4e10000000a42189b/content.htm?frameset=/en/46/
8c6361e4c70ad3e10000000a11466f/frameset.htm)

3.2   InfoCube

An InfoCube is a fact table and its associated dimension tables in the star schema.

3.3   DSO

DSO is the short version of DataStore Object.

A DataStore object serves as a storage location for consolidated and cleansed transaction data or master data on a document (atomic) level. This data can be evaluated using queries.

A DataStore object contains key fields (such as document number, document item) and data fields that, in addition to key figures, can also contain character fields (such as order status, customer).

In contrast to InfoCubes, the data in DataStore objects is stored in transparent, flat database tables. Fact tables or dimension tables are not created.

The DSO has typically been used as the store for incoming data which is further processed into the reporting InfoCubes. Therefore DSOs are the core building element of a layered scalable warehousing architecture (LSA). However with new technology many reporting scenarios can be now directly implemented on top of DSOs with good performance making InfoCubes completely obsolete in such cases (see also below in chapter "TCO reduction").

3.4   Multi-cube/Multi-provider

A multi-provider is a union of basic InfoProviders. The multi-provider itself does not contain any data; rather, data reside in the basic InfoCubes or DSOs. To a user, the multi-provider is similar to a basic InfoProvider. When creating a query, the user can select characteristics and key figures from different basic InfoProviders.[7]

4   The BW-EML Standard Application Benchmark

Among the SAP Standard Application Benchmarks the topic of analyzing large data volumes is also present for a long time now. Customers interested in information about performance and scalability of their SAP Business Warehouse implementations, initially turned to the BW benchmark where first results have been published 2003. This benchmark, like all SAP Standard Application Benchmarks was designed to represent relevant, real-life scenarios involving various SAP business applications to help customers find appropriate hardware configurations. Like all SAP standard application benchmark the execution just requires the complete SAP BW software stack and an additional benchmark package to run this benchmark. Because of the abstraction of the SAP stack of the DB provider and hardware this benchmark is open for many platforms. The first version of the BW benchmark consists of two separate phases one for uploading data and a separate phase for query processing. Both benchmark phases had individual metrics. For the load phase the throughput in terms of number of rows loaded per hour either to Info Cubes or Operational Data Store (ODS) was published, whereas for the query phase the number of query navigation steps per hour was mentioned. While initially this benchmark provided valuable results the need of more instant availability of business data in many use cases was recognized. In 2008 this lead to the next step of the evolution of SAP's benchmarks for the SAP Business Warehouse, the mixed load (BI-MXL) benchmark, which integrates the data upload and the query work load into a single benchmark phase. In this benchmark only the query navigation steps per hour were published. The data upload ran just as an additional background work load. The amount of data being uploaded during the benchmark phase of 1 h – 1/1000 of the original data in the system – represents the capability of the system to manage the near real time upload of data from the source system while handling heavy query work load.

With the changing demands of business decision making, like the instant availability of the latest business data and ad hoc query capability, the aspects that customers looked at when considering which configuration was the right fit changed as well. There was a clear need for a new benchmark type that covered these new

---

[7] http://wiki.scn.sap.com/wiki/display/BI/BW+Glossary,+Definitions+and+Explanations.

demands. SAP decided 2012 to develop a new set of metrics: the BW enhanced mixed load (BW-EML) benchmark.[8]

Today, when it comes to decision making systems, customers are looking for ones that support the following:

4.1 Near Real-Time Reporting Capabilities

To make informed decisions in a fast-paced world, the ability to get instant information from analytical business applications is crucial. Not only do businesses require quick access to information, they also need this information to include up-to-the minute details. Smart meter analytics and trade promotion management are just two examples of business processes that rely on near real-time reporting. To reflect this need we increased the frequency of the data upload compared to the BI MXL benchmark. Instead of having a total of three upload phases - one every 20 min, we increased this to a total of 12 delta uploads - one every 5 min.

4.2 Ad Hoc Reporting Capabilities

Data volumes in enterprise data warehouses have grown significantly over the past few years due to increased complexity of business data models and the level of detail captured in data. Sheer data volumes and the demand for unforeseeable navigation patterns make it impossible to use standard techniques like pre-aggregation of data for speeding up query response times. Modern analytical applications must allow users to navigate instantly through these huge amounts of data by providing extensive slicing and dicing functionality. The new benchmark reflects this need by using randomized sets of input parameters for the initial report execution and randomized drill down and filtering dimensions during query navigation steps.

Another alternative approach to show the ad hoc reporting capabilities would be to use more static queries and to combine this with disallowing any technique to pre-calculate and store final or intermediate query results. We decided explicitly not to use such an approach for several reasons.

Excluding specific technologies always has the big problem to define exactly what has to be excluded. There would always be the tendency to invent technologies which achieve similar results, but circumvent the definitions of things that are excluded.

Technology advances can lead to the fact, that the excluded technologies can still be applied in a beneficial way for the given use case. This might lead to an artificial constraint, so that not the best possible technology is used.

Verification of the benchmark results would need a procedure to verify, if a technique has been used which is not allowed. This would imply a deep knowledge of the used technologies which sometimes can even collide with the interest of the technology provider to protect their IP.

---

[8] http://global1.sap.com/campaigns/benchmark/appbm_bweml.epx.

Bottom line a benchmark should always test the capabilities of a system from the system usage point of view. How and with which technology a specific capability is achieved should not be important.

4.3 TCO Reduction

Since data warehouses can contain hundreds of terabytes of data, it is crucial to minimize data redundancy, while at the same time maintaining layered data models. With SAP NetWeaver Business Warehouse 7.30, it is possible to run reports directly on DataStore Objects (DSOs), which helps reduce TCO by saving precious storage space. DSOs are the core building element of a layered scalable warehousing architecture (LSA). Since reports can now analyze data in DSOs as fast as in multidimensional InfoCube data structures, InfoCubes have become completely obsolete in many reporting scenarios. To prove this capability we implemented the same set of queries with the same flexibility of query navigation steps on both – Info Cubes and DSOs.

4.4 The Main Differences Between BI-MXL and BW-EML

The Loading of delta requests happens every 20 min in BI-MXL. In BW-EML this timeframe was reduced to 5 min.

In BI-MXL there are pre-defined navigation paths and drilldowns only with the same characteristics set in display. In the BW-EML on the other hand ad hoc reporting requirements are verified by using randomized navigation paths and changes in the displayed characteristics.

All benchmark queries in the BI-MXL benchmark are defined on InfoCubes. The BW-EML benchmark uses DataStore Objects (flat table) and InfoCubes (star schema like data model) for reporting for a reduction of TCO.

The single metric of query navigation steps as well as the total data volume for the delta upload was kept like in the BW-MXL.

4.5 The BW-EML Benchmark in Detail

The BW-EML benchmark, like its predecessor, simulates a mix of multi-user reporting workload and the loading of delta data into the database simultaneously with user queries. Let's drill down further into the details of the benchmark definition.

4.6 Data Model

Even though in many cases InfoCubes do not provide a major benefit we like to insure that the database being benchmarked can efficiently use both InfoCubes and DSOs for reporting. Therefore the BW-EML benchmark's data model consists of three InfoCubes and seven DSOs, each of which contain the data of one specific year. The three InfoCubes contain the same data (from the last three years) as the corresponding DSOs.

Both types of objects consist of the same set of fields. The InfoCubes come with a full set of 16 dimensions, which comprise a total of 63 characteristics, with cardinalities of up to one million different values and one complex hierarchy. To simulate typical customer data models, the InfoCube is made up of 30 different key figures, including those that require exception aggregation. In the data model of the DSOs, the high cardinality characteristics are defined as key members, while other characteristics are modeled as part of the data members.

4.7   Scaling of Data Volumes

To test hardware configurations of various sizes, the BW-EML benchmark can be executed with different data volumes. The smallest configuration defined in the benchmark rules starts with an initial load of 500 million records (50 million records for each InfoCube and DSO). The records are loaded from ASCII flat files with a total record length of 873 bytes each. Larger volume configurations of the EML Benchmark include initial load volumes of one billion, two billion or even more records.

In each of the mentioned configurations, the total number of records that need to be loaded, in addition to the initial load during the benchmark run, is one thousandth of the initial amount of records. The high load phase of the benchmark must run for at least one hour. During this time, the delta data must be loaded in intervals of five minutes. The same number of records must be loaded in each InfoCube and DSO.

4.8   Query Model

For the BW-EML benchmark, eight reports have been defined on two MultiProviders – one for the three InfoCubes, and another for the seven DSOs. The respective reports on both MultiProviders are identical. This leads to two sets of four reports each. The four reports are categorized as follows:

- Report Q001: Customer-based reporting
- Report Q002: Material-based reporting
- Report Q003: Sales area-based reporting
- Report Q004: Price comparison reporting

During the report execution data is randomly selected for one particular year, and by that implicitly picking the InfoCube or DSO that contains the data. Up to six further navigation steps – filter and drill down steps with randomly chosen filter criteria and drill down dimension - are performed within each report, each of which results in an individual query and database access point.

Although the first three reports share similar navigation patterns, the filter and drill-down operations are randomized to address the demand for ad hoc types of queries. To make sure that the benchmark accesses different partitions of data, random choice of filter characteristics and the corresponding parameter values are used. Additionally, a random choice of characteristics for drilldowns or other slice-and-dice operations ensures that a huge number of different characteristic combinations are covered in a multi-user reporting scenario.

4.9   Multi-user Workload

A script controls the multi-user reporting workload in the BW-EML benchmark. The script starts for each user a simulated HTTP client which serves as a simulated frontend. By this not only the data base is tested but the complete backend stack including the SAP BW OLAP engine and the HTML rendering. With the script, the number of simulated users can be defined. By this the intended query throughput can be easily scaled through an increased number of simulated users. Each simulated user logs on to the system and then executes a sequence of four reports on InfoCubes and four reports on ODS objects including the ad hoc navigation steps consecutively, resulting in a total of 40 ad hoc navigation steps per user loop. After finishing all the predefined steps, the user logs off and then starts the next loop with a new log on. Simulated users are ramped up at a pace of one user logon per second. Once all configured users are running the high load phase starts and the benchmark control environment automatically starts a process chain that controls the delta load, which is scheduled every five minutes. The high load phase runs at least one hour. Within that hour each user runs multiple loops of the same sequence of 40 navigation steps. At the same time 12 delta uploads are started.

After a high load phase of at least one hour, the simulated users are ramped down, and the delta loads finish. A control program checks if the correct number of records has been uploaded during the benchmark run and if the uploaded records are visible in the report results. The essential key figure that is reported for a benchmark run is the number of ad hoc navigation steps per hour that the system executes successfully during the high load phase.

4.10   Verification of the Benchmark Results

The benchmark script environment automatically collects a large set of monitoring data available in the SAP system. This includes data like the number of table accesses to each individual table.

Specifically for the BW-EML (and BI-MXL) we collect statistical data for each individual query. This includes a total run time and the number of records processed from the Info Cubes or DSOs, This allows us to compare the query work load with previous runs to see if there are any unusual patterns visible. On top we implemented a test query which verifies after each delta upload, if the data has been loaded correctly and in time.

4.11   Key Challenge for the Design of the BW-EML Benchmark

One the one hand benchmarks need to provide highly reproducible results to be relevant, because otherwise results cannot be compared. On the other hand to reflect the ad hoc query capabilities we decided to introduce a huge variability into the query work load, with a huge amount of possible different query navigations and drilldowns.

Two things are important to nevertheless achieve good reproducibility

1. Characteristics are grouped by their respective cardinalities, and only characteristics of the same cardinality are considered for a randomized operation. E.g. we have a group of high cardinality characteristics like customer and material with 1 million distinct values and groups of characteristics with 1000 like e.g. plant, with 100 like sales organization or with 10 distinct values like distribution channel, Even though the navigation path is chosen randomly, the data volume that needs to be processed is similar independent of the chosen navigation path.

   Because of the data skew that is also included in the data generator the drill down to the most detailed level of data nevertheless leads to a significant different amount of processed data.

   Data skew is introduced for characteristics which are obviously depended on each other in most cases. E.g. there is a different characteristic for "ship to party" and the "bill to party". In real life in most cases they are identical. Here we simulated it, by setting the same value in 95 % of the generated records. Another type of data skew introduced was sparsity, so that only a small fraction of the records contained a different value and all other records either contained empty values or one specific default value.

   This data skew led to the fact that the actual processed data varied from a few result records up to a couple of 100 thousand resulting records.

   Even with these significant differences in the execution of equivalent steps – so same step number in different loops – we nevertheless could achieve well reproducible results. Actually here the second effect helped us.

2. Even for the smallest benchmark the total number of more than 60,000 queries executed during the benchmark run leads to a sufficient averaging of the individual queries so that the final benchmark result is well reproducible. We never did an in depth statistical analysis, but looking at the results of run times and records processed of individual queries in detail it's obvious that for each navigation step only a limited number of variations of an individual step seemed to be possible.

5 Results of the Benchmarks

So far the following 9 results have been published for the SAP BW-EML standard application benchmark Fig. 3. [9,10]

The result of a series of identical performance test runs using the BW-EML workload was that the variability between the different runs is less than 1 %. The results were measured internally at SAP on the SAP HANA database Fig. 4.

---

[9] http://global.sap.com/solutions/benchmark/bweml-results.htm.

[10] http://global.sap.com/campaigns/benchmark/appbm_cloud.epx.

Certifi-cation	Number of initial records	Ad-hoc navigation steps per hour	Scale-out	Number of DB Servers	Number of Application Servers
2013020	500,000,000	66,900	No	1	1
2014001	500,000,000	113,390	No	1	2
2014013	500,000,000	137,510	No	1	2
2012023	1,000,000,000	65,990	No	1	1
2013027	1,000,000,000	129,930	No	1	1
2014021	2,000,000,000	111,850	No	1	2
2014009	2,000,000,000	126,980	No	1	2
2014014	2,000,000,000	177,590	Yes	5	3
2013037	3,000,000,000	128,650	Yes	4	2

Fig. 3. Overview SAP BW-EML standard application benchmark results

Run Number	Date	Time	Ad-hoc navigation steps per hour
1	02/22/13	09:04:29	62008
2	02/22/13	10:09:36	62311
3	02/22/13	11:14:15	62059
4	02/22/13	12:19:05	62412
5	02/22/13	13:24:12	62363

Fig. 4. Table of performance test results BW-EML

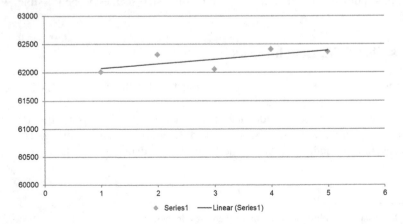

Fig. 5. Chart of performance test results BW-EML

What can be seen from the certified benchmark results and the internal testing performed so far is that the challenge of the benchmark was mastered. The BW-EML benchmark provides highly reproducible results and at the same time delivers the ad hoc query capabilities with a wide variability of the query work load and a huge amount of different query navigations and drill downs (Fig. 5).

6   Conclusion

Compared to earlier benchmarks the main difference is the usage of random query navigation steps and random choice of initial filter characteristics with a huge number of possible variations. This leads to a query work load where pre-calculated query results and aggregates are of limited value and so this benchmark stresses the ad hoc query capability to a large extend.

Also we have shown that it is possible to design a benchmark work load which on the one hand introduces a huge variability into the work load by choosing a data generator which generates random values including data skew, by using random query navigation steps and random query predicates and on the other hand achieve well reproducible results.

Benchmarking Big Data Systems: Introducing TPC Express Benchmark HS

Raghunath Nambiar[✉]

Cisco Systems, Inc., 275 East Tasman Drive,
San Jose, CA 95134, USA
rnambiar@cisco.com

Abstract. As Big Data becomes an integral part of enterprise IT ecosystem, many organizations are challenged with metrics and methodologies to benchmark various hardware and software technologies in terms of performance, price-performance and energy efficiency. Continuing its commitment to bring relevant benchmarks to industry, the Transaction Processing Performance (TPC) formed the Big Data committee to develop set of industry standards, and announced its first standard, the TPC Express benchmark HS (TPCx-HS) in August 2014. TPCx-HS enables measurement of both hardware and software including Hadoop Runtime, Hadoop file system API compatible systems and MapReduce layers. The TPCx-HS can be used to asses a broad range of system topologies and implementation methodologies, in a technically rigorous and directly comparable, vendor-neutral manner.

Keywords: TPC · Big data · Benchmark · Hadoop

1 Introduction

Over the past quarter century, industry standard benchmarks from Transaction Processing Performance (TPC) and Standard Performance Evaluation Corporation (SPEC) have had a significant impact on the computing industry. Many buyers use the results as points of comparison when evaluating new computing systems. Vendors use the standards to illustrate performance competitiveness for their existing products, and use internally to improve and monitor the performance of their products under development. Industry standard benchmarks are widely used by research communities as well [1, 2].

Founded in 1988, the Transaction Processing Performance Council (TPC) has the history and experience to create industry standard benchmarks. Continuing on the Transaction Processing Performance Council's commitment to bring relevant benchmarks to industry, the TPC developed TPC Benchmark HS (TPCx-HS) - the first standard that provides verifiable performance, price-performance and energy consumption metrics for big data systems. The formation of TPC Big Data committee and development of TPCx-HS were influenced by the Technology Conference on Performance Evaluation and Benchmarking (TPCTC) and the Workshops on Big Data Benchmarking (WBDB) [3, 4, 5].

© Springer International Publishing Switzerland 2015
T. Rabl et al. (Eds.): WBDB 2014, LNCS 8991, pp. 24–28, 2015.
DOI: 10.1007/978-3-319-20233-4_3

The TPCx-HS can be used to asses a broad range of system topologies and implementation methodologies, in a technically rigorous and directly comparable, vendor-neutral manner. And while modeling is based on a simple application, the results are highly relevant to Big Data hardware and software systems [6].

2 TPC

The TPC is a non-profit corporation founded to define transaction processing and database benchmarks and to disseminate objective, verifiable TPC performance data to the industry. Over the years, TPC benchmarks have raised the bar for what the computing industry has come to expect in terms of benchmarks themselves.

Though the original focus has been on online transaction processing (OLTP) benchmarks, to-date the TPC has approved a total of twelve independent benchmarks. Of these benchmarks, TPC-C, TPC-H, and TPC-E, TPC-DS, TPC-DI, TPC-V and the new TPCx-HS are currently active. TPC Pricing Specification and TPC Energy Specification are common across all benchmark standards designed to allow a fair comparison for pricing and energy efficiency.

To address the increased benchmark development cycle and development costs, in 2103, the TPC created new benchmark development track called the "express benchmark", easy to implement, run and publish, and less expensive and available online via downloadable kit. TPCx-HS is the first TPC Express benchmark [7, 8]. See TPC Benchmark Timeline in Table 1.

Table 1. TPC benchmark timeline

Benchmark Standards	1988	1989	1990	1991	1992	1993	1994	1995	1996	1997	1998	1999	2000	2001	2002	2003	2004	2005	2006	2007	2008	2009	2010	2011	2012	2013	2014
TPC-A		■	■	■	■	■	■	■																			
TPC-B			■	■	■	■	■	■																			
TPC-C					■	■	■	■	■	■	■	■	■	■	■	■	■	■	■	■	■	■	■	■	■	■	■
TPC-D								■	■	■	■																
TPC-R											■	■															
TPC-H												■	■	■	■	■	■	■	■	■	■	■	■	■	■	■	■
TPC-W													■	■	■												
TPC-App																	■	■									
TPC-E																			■	■	■	■	■	■	■	■	■
TPC-DS																									■	■	■
TPC-VMS																									■		
TPC-DI																										■	
TPCx-HS																											■
TPC Express Benchmarks																											
Common Specifications																											
Pricing																		■	■	■	■	■	■	■	■	■	■
Energy																											
Developments in Progress																											
TPC-VMC																						■	■	■	■	■	■
TPC-V																											

3 TPCx-HS

TPCx-HS was developed to provide an objective measure of hardware, operating system and commercial Apache Hadoop File System API compatible software distributions, and to provide the industry with verifiable performance, price-performance and availability metrics. The benchmark models a continuous system availability of 24 h a day, 7 days a week [6].

Even though the modeled application is simple, the results are highly relevant to hardware and software dealing with Big Data systems in general. The TPCx-HS stresses both hardware and software including Hadoop run-time, Hadoop Filesystem API compatible systems and MapReduce layers. This workload can be used to asses a broad range of system topologies and implementation of Hadoop clusters. The TPCx-HS can be used to asses a broad range of system topologies and implementation methodologies in a technically rigorous and directly comparable, in a vendor-neutral manner.

The main components of the TPCx-HS are detailed below:

Workload: TPCx-HS is based popular TereSort, and workload consists of the following modules:

- HSGen is a program to generate the data at a particular Scale Factor. HSGen is based on TeraGen.
- HSDataCheck is a program to check the compliance of the dataset and replication.
- HSSort ia program to sort the data into a total order. HSSort is based on TeraSort.
- HSValidate is a program that validates the output is sorted. HSValidate is based on TeraValidate.

The benchmark test consists of two runs, and the slower run is reported.

Scale Factor: The TPCx-HS follows a stepped size model. Scale factor (SF) used for the test dataset must be chosen from the set of fixed Scale Factors defined as follows:

- 1 TB, 3 TB, 10 TB, 30 TB, 100 TB, 300 TB, 1000 TB, 3000 TB, 10000 TB.
- The corresponding number of records are 10B, 30B, 100B, 300B, 1000B, 3000B, 10000B, 30000B, 100000B, where each record is 100 bytes generated by HSGen.

The TPC will continuously evaluate adding larger Scale Factors and retiring smaller Scale Factors based on industry trends.

A TPCx-HS Result is only comparable with other TPCx-HS Results of the same Scale Factor. Results at the different Scale Factors are not comparable, due to the substantially different computational challenges found at different data volumes. Similarly, the system price/performance may not scale down linearly with a decrease in dataset size due to configuration changes required by changes in dataset size.

Metrics: TPC-xHS defines three primary metrics.

The performance metric of the benchmark is HSph@SF, the effective sort throughput of the benchmarked configuration:

$$HSph@SF = \frac{SF}{T/3600}$$

Where:

SF is the Scale Factor

T is the total elapsed time for the run in seconds

The price-performance metric for the benchmark is defined as:

$$\$/HSph@SF = \frac{P}{HSph@SF}$$

Where:

P is the total cost of ownership of the SUT.

System Availability Date

The System Availability Date is when then benchmarked systems are generally available.

The TPCx-HS energy results are expected to be accurate representations of system performance and energy consumption. Therefore there are certain requirements which must be followed. The approach and methodology are explicitly detailed in this specification and the TPC Benchmark Standards, as defined in TPC-Energy.

4 TPCx-HS Benefits

Benefits of TPCx-HS benchmark to end-users and vendors benefits are outlined below:

- Enable direct comparison of price, price-performance and energy efficiency of different vendor architectures. The standard requires vendors to specify their hardware and software components and to disclose general availability, cost of ownership, and maintenance fees for three years with a continuous system availability of 24 h a day, 7 days a week.
- Independent audit by a TPC-certified auditor or peer review committee for compliance before vendors can publish benchmark results. Further any TPC member can challenge results for up to sixty days.
- Full disclosure report sufficient to allow interested parties to evaluate and, if necessary, recreate the implementation of the benchmark. Full disclosure report is publically available at www.tpc.org
- Significantly lower cost to the vendors as full kit is developed and provided by the TPC. The kit includes the specification document, users guide documentation, shell scripts to set up the benchmark environment and Java code to execute the benchmark load.

5 Conclusion

TPCx-HS is the first major step in creating a set of industry strands for measuring varies aspects of hardware and software systems dealing with Big Data. The author envision that TPCx-HS will be a useful benchmark standard to buyers, as they evaluate new systems for Hadoop deployments in terms of performance, price-performance and energy efficiency. Also, enabling healthy competition between vendors that will result in product developments and improvements.

Acknowledgement. Developing an industry standard benchmark for a new environment like Big Data has taken the dedicated efforts of experts across many companies. The author thank the contributions of Andrew Bond (Red Hat), Andrew Masland (NEC), Avik Dey (Intel), Brian Caufield (IBM), Chaitanya Baru (SDSC), Da Qi Ren (Huawei), Dileep Kumar (Cloudera), Jamie Reding (Microsoft), John Poelman (IBM), Karthik Kulkarni (Cisco), Meikel Poess (Oracle), Mike Brey (Oracle), Mike Crocker (SAP), Paul Cao (HP), Reza Taheri (VMware), Simon Harris (IBM), Tariq Magdon-Ismail (VMware), Wayne Smith (Intel), Yanpei Chen (Cloudera), Michael Majdalany (L&M), Forrest Carman (Owen Media) and Andreas Hotea (Hotea Solutions).

References

1. Nambiar, R., Lanken, M., Wakou, N., Carman, F., Majdalany, M.: Transaction processing performance council (TPC): twenty years later – a look back, a look ahead. In: Nambiar, R., Poess, M. (eds.) TPCTC 2009. LNCS, vol. 5895, pp. 1–10. Springer, Heidelberg (2009)
2. Nambiar, R., Wakou, N., Carman, F., Majdalany, M.: Transaction processing performance council (TPC): state of the council 2010. In: Nambiar, R., Poess, M. (eds.) TPCTC 2010. LNCS, vol. 6417, pp. 1–9. Springer, Heidelberg (2011)
3. Workshops on Big Data Benchmarking (WBDB): http://clds.sdsc.edu/bdbc
4. Baru, C., Bhandarkar, M., Nambiar, R., Poess, M., Rabl, T.: Setting the direction for big data benchmark standards. In: Nambiar, R., Poess, M. (eds.) TPCTC 2012. LNCS, vol. 7755, pp. 197–208. Springer, Heidelberg (2013)
5. TPC forms Big Data Benchmark Standards Committee: http://blogs.cisco.com/datacenter/tpc-bd/
6. TPCx-HS Specification: www.tpc.org/tpcx-hs
7. Huppler, K., Johnson, D.: TPC express – a new path for TPC benchmarks. In: Nambiar, R., Poess, M. (eds.) TPCTC 2013. LNCS, vol. 8391, pp. 48–60. Springer, Heidelberg (2014)
8. Nambiar, R., Poess, M.: Keeping the TPC relevant! PVLDB **6**(11), 1186–1187 (2013)

Benchmarking *Internet of Things* Solutions

Ashok Joshi[1(✉)], Raghunath Nambiar[2], and Michael Brey[1]

[1] Oracle, Redwood shores, USA
{ashok.joshi,michael.brey}@oracle.com
[2] Cisco, San Francisco, USA
rnambiar@cisco.com

Abstract. The Internet of Things (IoT) is the network of physical objects accessed through the Internet, as defined by technology analysts and visionaries. These objects contain embedded technology allowing them to interact with the external environment. In other words, when objects can sense and communicate, it changes how and where decisions are made, and who makes them. In the coming years, the Internet of Things is expected to be much larger than the internet and world-wide web that we know today.

1 Introduction

The *Internet of Things* (IoT) promises to revolutionize various aspects of our lives. Until recently, most of the internet activity has involved human interaction in one way or another, such as web browser-based interaction, e-commerce, smart-phone apps and so on. Recently, there has been a lot of interest in connecting devices of various sorts to the internet in order to create the "internet of things" for delivering dramatic benefits relative to existing solutions. For example, network-enabled home health-care monitoring devices hold the promise of improving the quality of healthcare while reducing costs. Connected energy monitoring systems enable more efficient utilization of energy resources and reduce global warming. Traffic management systems aim to reduce congestion and decrease automobile accidents. It is clear that such systems can deliver tremendous value and benefits over traditional solutions.

IoT has fueled massive investment and growth across the entire spectrum of industry and public sectors. Companies such as Oracle [1], Cisco [2], Intel [3], Microsoft [4] and others, are investing heavily in IoT initiatives in order to address the computing and infrastructure requirements of this market.

Though the IoT industry is still in its infancy, based on the number and variety of solutions being proposed, it seems clear that IoT solutions will play a very significant role in the coming years. Although there has been a lot of IoT marketing and promotional activity from small and large vendors alike, there is a paucity of proposals to measure the performance of such systems. At best, some of the proposals [5, 6, 7] either measure the performance of a specific component and extrapolate the results to a broader

---

This paper proposes an approach to benchmarking the server-side components of a complete Internet of Things solution.

© Springer International Publishing Switzerland 2015
T. Rabl et al. (Eds.): WBDB 2014, LNCS 8991, pp. 29–36, 2015.
DOI: 10.1007/978-3-319-20233-4_4

IoT solution [5], or propose a very broad and comprehensive benchmark *framework* that encompasses performance, public policy and socio-economic considerations [7].

In contrast, our approach is much more specific and focuses exclusively on the performance metrics of an IoT system (in particular, on the server-side components of an IoT solution) using a representative workload. We believe that a systematic approach to performance benchmarking of IoT systems will help users evaluate alternative approaches, and ultimately accelerate adoption; in other words, our approach is workload and solution-centric. A well-crafted performance benchmark also serves as a catalyst for vendors to improve the performance as well as the overall usability of the solution, which should benefit the IoT consumer as well.

In the rest of the paper, we describe the components of a typical IoT solution, then describe a representative use-case, followed by a straw-man description of a benchmark intended to measure the performance (throughput, latency, etc.) of such a system. Admittedly, this is a straw-man proposal; we explore how this benchmark might be applicable to other use-cases (with modifications) and also describe some of the limitations of our proposal and areas for future work. Significant additional work is necessary in order to create one or more formal benchmarks for IoT systems.

2   IoT Solution Architecture Components

Although the scale of IoT solutions is enormous in comparison to current internet usage, in this paper, we address only the essential architecture of an IoT solution in the context of benchmarking the server components. The following discussion presents a simplified view of the system as a small number of distinct, essential components. These architecture components and their relationships are illustrated in Fig. 1 and described below in more detail.

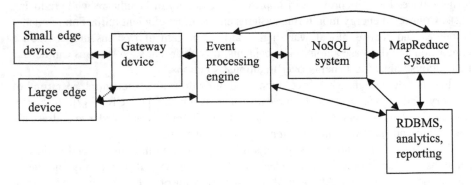

Fig. 1 Essential components of a typical IoT architecture.

At the "device" edge of the network, we have *edge devices* and software that controls the devices. Quite often, the devices are relatively inexpensive and have limited computing and network capabilities (*small edge devices*) – temperature sensors in office buildings, various sensors in a vehicle, blood pressure and blood glucose monitors are all examples of such devices. Mobile phones, computers in vehicles are also examples of edge devices which have better computing and local storage capabilities than small edge devices. Small edge devices are primarily intended to capture and send information from the object being monitored, to the central repository; secondarily, they may also receive data to make appropriate changes (e.g. a building temperature sensor might receive a directive to shutdown the air conditioning system in the event of a fire). More capable edge devices like mobile phones (large edge devices) typically provide a richer set of capabilities beyond simple monitoring, and can have bi-directional exchange of information with the central data repository.

Depending on the number and capability of edge devices in the deployed topology, edge devices (such as sensors) connect to a *gateway device*, organized as a multi-level hierarchy, with a set of small devices being served by a gateway device. The gateway device with sufficient computing, storage and network connectivity is essentially a "concentrator" that can process data and exchange information between the small devices and the data center. A gateway device communicates with a relatively small set of edge devices (typically less than one hundred) on one side and the data center on the other. For example, a set-top box (gateway device) might serve as the connectivity "hub" for all the sensors and devices in a home. Or a laptop/desktop-class device might serve as the connectivity hub for sensors on a set of machines and robots on a factory floor. A gateway device is connected to the edge devices on one side, and to the data center (or other gateway devices in a multi-stage hierarchy) on the other side and enables efficient communication and data exchange between edge devices and the rest of the system. Note that in a large system, there might be multiple gateway stages, in order to simplify data management and communication in the system.

A large edge device may either be connected directly to the central repository, or indirectly through an intermediate gateway device, depending on the specific application scenario. For example, a mobile phone might connect directly with the data center, without any intermediate gateway device.

The other major component in the hierarchy is obviously the network. The network is not only responsible for communication, but it also facilitates or enforces attributes such as security, privacy, reliability, availability and high performance.

Mobile edge devices pose an interesting challenge since they may occasionally be out of range of a network connection. For example, an edge device mounted on a delivery truck may occasionally lose network connectivity when the truck enters a tunnel or is in a remote area. In order to avoid data loss while the device is disconnected from the network, it is necessary to temporarily store new data in the (large) edge device and send that information to the data center when connectivity is restored. Even in cases where temporary loss of network connectivity is not an issue, it may still be valuable to store some data locally on the edge device or gateway device and only synchronize it periodically with the data center, in order to minimize costs associated with network usage, since these costs can add up quickly, especially at IoT scale. Oracle Database Mobile Server [8] is an example of a technology that is well suited for data synchronization between edge device repositories and the data center. Lightweight, embeddable databases such as SQLite [9], Oracle Berkeley DB [10] and Java DB [11] are well suited for reliable, device-local data management.

Finally, the *data center* (centralized or distributed) receives data from the edge devices and processes it in various ways, depending on the application. Broadly speaking, data center processing can be classified into *event processing, high throughput data capture, device management, data aggregation, data analytics* and *reporting*.

We will use the term *event processing* to describe the processing that is required to respond to an event as quickly as possible. For example, if a sensor in a building sends information about a potential fire, then it is necessary to respond to that message as quickly as possible. Event processing typically involves various lookups (e.g. look up the location of the device, determine location of the closest emergency responders) as well as generating appropriate responses (e.g. send a message back to the sensor to switch on the sprinkler system, send an alert to first responders). Note that of all the messages being received from the edge devices, only a small fraction need an immediate response as described above. However, all the messages must be examined by the event processing engine in order to determine whether an immediate response is required.

Messages from edge and gateway devices need to be captured persistently in order to enable further analytics. Though message sizes are generally small, typically varying from few tens of bytes, to a few kilobytes, the sheer volume of devices involved and the rate of data capture usually implies that a scalable, persistent store to capture the data is required. Depending on the nature of data and the application, a NoSQL database system or a distributed file system such as HDFS might be used. For example, data related to medical information (e.g. blood pressure, heart rate and blood glucose readings) for patient monitoring might need to be stored in a NoSQL database for reliability and regulatory reasons. On the other hand, a distributed file system such as HDFS might be sufficient to store messages containing information from office building room temperature sensors.

A scalable database system (e.g. a NoSQL system) is also typically required to manage information about the edge and gateway devices. Such information typically includes the ID and owner of the device, location, model number and other details, software patch version etc. Occasionally, it is necessary to send updates or new information to the devices, for example, to deliver software upgrades, or change configuration settings. In an IoT environment, it is very likely that such information needs to be managed for hundreds of thousands to millions of edge devices; a NoSQL repository is ideally suited for this purpose.

Lastly, the data received from the devices needs to be aggregated and processed in various ways to identify trends, opportunities for optimization as well as to generate reports required for running the business. For example, data collected from vehicle brake sensors might be correlated with GPS locations and time-of-day in order to optimize delivery routes. Data from heart rate sensors and physical activity sensors might be used in order to recommend an appropriate exercise regimen for heart disease patients.

3   Benchmarking IoT

IoT is rapidly emerging as the "next generation" technology, and promises to revolutionize and improve various aspects of everyday life. There are literally hundreds of commercial and open source organizations in this broad and somewhat amorphous space.

Regardless of the kinds of solutions available, we believe that a systematic approach to measurement and benchmarking such systems will provide significant benefits to the user community and solution providers.

Attempting to define a comprehensive benchmark that deals with the various components in the end-to-end solution is a daunting task; in the following paragraphs, we describe a skeleton benchmark intended to address performance issues of some of the server-side components. The suggestions and recommendations outlined below have been derived from the authors' real-world understanding and experience with early deployments of IoT solutions. In the future, this work can be extended to define additional benchmarks to address other aspects of an IoT solution.

The proposed "strawman" benchmark is intended to model and measure some of the server-side components of a real-world scenario viz. to capture data from home health-care monitoring systems for millions of patients. This involves capturing health-related readings reliably and securely from large numbers of devices, responding rapidly to abnormal events, and analyzing vast amounts of data in order to benefit the patient community.

A *reading* from an edge device is typically short (tens of bytes) and contains information like device ID, timestamp, geo-location (if appropriate) and measurements of the variables that the sensor captures. For the purposes of a benchmark, a reading is 96 bytes of data, with a 16 byte key (device ID) and five 16-byte fields of data. Let us assume that each edge device (e.g. a heart rate monitor) captures a reading every 10 min. For a population of one million patients, this translates to a reading capture rate of 1667 readings/sec and 160 KB/sec. As soon as the reading is available, the edge device sends the data to the data center either directly or via intermediate gateway devices.

An event processing system processes every incoming reading. A small percentage (e.g. 1 percent) of the incoming readings will need a response immediately and every reading needs to be saved in a secure, persistent repository. The event processing system interprets the incoming reading and decides whether it requires an immediate response. This decision is based on the reading itself as well as information about the specific patient (this requires lookups in the patient database). For the purposes of the benchmark, let us assume that the decision involves looking up a single database that maintains per-patient information and a history of the recent three months' worth of heart-rate readings (about 13000 heart-rate readings, or 1.2 MB as well as the patient's recent medical history of 100 KB).

Regardless of whether they require an immediate response or not, all readings are stored in the patient history repository. Based on the previous description, for a patient population of one million patients, this translates to a repository that can manage approximately 1.3 TB of patient data.

Periodically, a batch job reads the heart rate readings history for each patient and consolidates multiple "similar" adjacent readings into a single reading. Since each sensor measures the heart-rate of the patient every 10 min, it is possible that a series of readings might have heart-rate readings that are within the normal range for the patient. For example, while the patient is asleep, the heart-rate might stay relatively constant for a significant interval of time. The batch job identifies such intervals and consolidates multiple readings at taken at 10 min intervals into a single "average" reading for a longer interval. For example, if the patient's heart-rate doesn't fluctuate by more than 3 heartbeats per minute between two subsequent readings, these two readings will be "consolidated" into a single reading which has the average value of the two original readings and the time interval will be changed to be a 20 min span. This consolidation operation has the dual benefit of reducing the amount of storage required and also simplifying the job of the physician by consolidating similar, adjacent readings into a single reading over a larger interval. For the purposes of the benchmark definition, we assume that 50 % of the readings are consolidated into a single reading. This translates to a repository of 1.3 TB of patient data for a patient population of 1 million patients.

Operation Profiles. The message format and size, as well as the patient record size and content have been described earlier. The various operation profiles are as follows:

- Incoming reading received by the event processing engine. For each reading, the event processing engine looks up the related patient record. For 1 % of the messages (selected randomly), it generates an alert. Every message is appended to the related patient record. From a benchmarking perspective, this operation is intended to model the data capture throughput and data retrieval throughput of the system.
- Consolidating similar, adjacent heart-rate readings. A background job reads each patient reading, consolidates similar, adjacent heart-rate readings into a single reading and updates the patient record appropriately. The choice of whether to retain the readings that contributed to the consolidated reading or delete such readings is left up to the implementers of the benchmark. The consolidation operation should be able to consolidate 50 % of the readings within a 24 h window. The batch job must retrieve every record in the database within the same period. From a benchmarking perspective, this operation is intended to model the analytics and batch-processing aspects of the system.

A key premise of most IoT systems is the ability to respond in a timely manner to abnormal events. Secondly, the volume of readings and measurements from sensors can be enormous. The ability to filter the "signal" from the "noise" is critical to efficient operation of the system. These operation profiles attempt to capture these aspects in a benchmark.

Performance and Price-Performance Metrics. The benchmark measures the throughput of data capture (readings/sec) in the patient repository under the constraints described earlier. Further, data capture, response generation and data consolidation jobs must be run concurrently, since the system is intended to run 24X7. It is difficult to propose a price-performance metric for the benchmark, since several necessary components of the complete system are intentionally ignored in this proposal.

Other Considerations. This paper provides some recommendations for benchmarking the event processing and data repository components of a IoT system intended to capture and process data for a patient data monitoring system. It also suggests scaling rules for such a benchmark.

Several variants of the above approach are possible, depending on the specific IoT application being modeled. It may be necessary to vary some of the parameters as follows:

- The number of edge devices, typical size of a message from each device and the frequency of sending a message from each edge device.
- Event processing requirements.
- Number of historical messages to be retained (in this proposal, we recommended storing three months worth of readings).
- Size of the device record (e.g. the size of each patient record, in this proposal).
- Other parameters as relevant to a specific usage scenario.

From a benchmarking perspective, this proposal addresses the following essential components:

- Data capture from multiple sources (edge or gateway devices) - this measures "raw" data capture throughput.
- "Real-time" analysis of the input data for fast response to a subset of the incoming data – this measures the performance of the event processing engine as well as the throughput and latency of accessing reference data from the data store.
- Consolidation of the incoming data in order to conserve space – this measures batch processing (as a proxy for analytics processing) performance of the system.

We believe that server-side data capture, real-time response, and batch processing performance are common to a wide variety of other IoT usage scenarios such as climate control, vehicle monitoring, traffic management and so on. With appropriate changes to model a particular scenario, this proposal can be adapted to measure and benchmark the performance of the server components of such scenarios.

4 Conclusion

Internet-of-Things (IoT) represents a broad class of emerging applications that promises to affect everyday life in various ways. The scale of IoT applications dwarfs the use of the internet, as we know it today. Many commercial and open source players are investing heavily in solutions intended to address specific aspects of IoT; some of the larger players claim to provide a comprehensive, end-to-end solution.

For all these scenarios, we believe that a set of benchmarks, modeled on real-world scenarios, will provide a valuable yardstick to compare the relative merits and costs of such solutions and will have a significant beneficial impact on the IoT industry. In this paper, we have attempted to provide some guidelines on how one might go about designing such benchmarks. Significant additional work is needed in order to define formal benchmarks; hopefully, this proposal will inspire some of that future work.

References

1. http://www.oracle.com/iot
2. http://www.cisco.com/web/solutions/trends/iot/overview.html
3. http://www.intel.com/iot
4. http://www.microsoft.com/windowsembedded/en-us/internet-of-things.aspx

5. http://www.sys-con.com/node/3178162
6. http://www.probe-it.eu/?page_id=1036
7. http://www.probe-it.eu/wp-content/uploads/2012/10/Probe-it-benchmarking-framework.pdf
8. http://www.oracle.com/technetwork/database/database-technologies/database-mobile-server/overview/index.html
9. http://www.sqlite.org
10. http://www.oracle.com/technetwork/database/database-technologies/berkeleydb/overview/index.html
11. http://www.oracle.com/technetwork/java/javadb/overview/index.html

Benchmarking Elastic Query Processing on Big Data

Dimitri Vorona[✉], Florian Funke, Alfons Kemper, and Thomas Neumann

Fakultät Für Informatik, Technische Universität München, Bolzmannstraße 3,
85648 Garching, Germany
{vorona,funkef,kemper,neumann}@in.tum.de

Abstract. Existing analytical query benchmarks, such as TPC-H, often assess database system performance on on-premises hardware installations. On the other hand, some benchmarks for cloud-based analytics deal with flexible infrastructure, but often focus on simpler queries and semi-structured data. With our benchmark draft we attempt to bridge the gap by challenging analytical platforms to answer complex queries on structured business data while leveraging the elastic infrastructure of the cloud to satisfy performance requirements.

1 Introduction

Modern cloud architectures constitute new opportunities and challenges for analytical platforms. They allow to dynamically adapt cluster sizes as client's performance, workload or storage requirements change over time. In contrast to the cloud infrastructure, the natural habitat of analytical relational database systems is typically a single machine or small-sized cluster of dedicated servers. Naturally, this is also reflected in today's analytical benchmarks that assess performance or price/performance of RDBMSs on such an infrastructure.

To leverage the full potential of cloud infrastructures, *elastic scaling* is crucial and currently rarely measured by existing analytical DBMS benchmarks. Elastic scaling describes the ability of a system to rapidly grow and shrink based on temporary changes in performance requirements.

While scaling elastically on such a dynamic infrastructure can be a challenge for relation database systems, cloud-native platforms such as MapReduce were build to operate on hundreds of nodes. However, they often cannot compete with relational systems in terms of performance, in particular when processing complex queries on structured business data [8,19]. This benchmark proposal attempts to challenge both classical RDBMSs as well as solutions specifically built to run in the cloud with elastic query processing. We attempt to assess how well analytic platforms perform in a typical OLAP scenario using flexible cluster-sizes. In contrast to existing benchmarks, tested systems must *adaptively* scale the number of nodes up and down, as depicted in Fig. 1. This simulates the business requirement to achieve both good performance and low costs.

The proposed benchmark targets a multitude of different systems, from classical RDBMSs to MapReduce-based systems such as Apache Hive. Contestants

© Springer International Publishing Switzerland 2015
T. Rabl et al. (Eds.): WBDB 2014, LNCS 8991, pp. 37–44, 2015.
DOI: 10.1007/978-3-319-20233-4_5

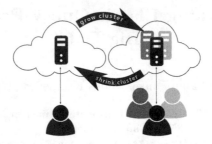

Fig. 1. Business Scenario: Grow cluster as workload grows to ensure performance. Shrink cluster when possible to save costs.

need be able to process SQL or a similar derivative (e.g. HiveQL). It is important to note that this benchmark does not attempt to compare features (e.g. schema-free vs predefined schema) or programming/query languages of different types of data management and data processing systems.

2   Related Work

A number of benchmark proposals aim to compare RDBMSs and NoSQL data stores with specific focus on scalability on modern distributed architectures. The Yahoo Cloud Serving Benchmark (YCSB) [6] is widely regarded as the leading benchmark in this area. While the YCSB describes a simple *transactional* workload rather than an analytical one, it specifies metrics applicable to a wide range of benchmarks for distributed systems.

The authors of [11] present a general framework to evaluate the costs of sub-optimal elasticity using predefined demand curves. In [7] the authors developed an elasticity measurement framework for cloud databases using a YCSB-like workload and custom system stability criteria. The work in [10] presents a method of estimating actuation delays in cloud systems, i.e. time between the configuration change command and the stabilization of the system's performance in the new state. The framework described in [14] is able to manage cluster size adjustment and reconfiguration depending on workloads. The authors also evaluate associated elasticity costs for node addition and removal.

While the aforementioned research tackles the challenge of assessing the elasticity of different DBMSs, the workloads are transactional, with simple key-value operations. An extensive comparison of distributed RDBMSs with other approaches like MapReduce for large-scale data *analysis* is presented in [19] and [22]. The focus of the experiments is on performance and resources utilization using data sets in the range of hundreds of terabytes as well as user-level aspects, e.g. ease of use. Elasticity is not measured. Similarly, [8] compared the performance of a parallel DBMS system with NoSQL-based solutions under different workloads, including decision support. Findings include that the tested RDBMS scales almost linearly and provids better performance than Hive, but the relative speed-up goes down as the scale factor increases.

Elasticity of RDBMSs is often questioned [3,6], however, recent developments improve this aspect significantly. ScyPer [16] is a main-memory DBMS which was designed to elastically scale out on cloud architectures. Related research in [21] shows that optimizing relational operators can lead to drastically improved resource utilization in distributed systems. AnalyticsDB [23] is a distributed analytical in-memory database management system based on RAMCloud [18]. The performance evaluation proves its scalability and elasticity using the Star Schema Benchmark [17], but does not allow direct comparisons with other systems. Additionally, the properties scalability and elasticity are neither clearly separated nor is there a single metric to quantify the elasticity.

3 Benchmark Design

The Elastic Query Processing Benchmark (EQP Benchmark) strives to assess the ability of analytical platforms to quickly and efficiently adapt to changing cluster sizes. To do so, it simulates an exploratory analysis of operational business data. The analysis is conducted at different scales to simulate typical patterns, e.g. diurnal [15], where the requirements increase gradually with the start of the business hours and fall at the end of the day. The benchmark design attempts to follow the criteria for Big Data benchmarks postulated in [4]. EQP Benchmark is defined on an abstract level and does not enforce specific technologies, so it can be used to compare elasticity of a wide range of systems.

Following the suggestion in [5], the proposed benchmark avoids making assumptions about the distributed computing architecture (e.g. shared nothing vs shared everything) or cluster organization. It also clearly differentiates between the two related system properties *scalability* and *elasticity*.

The EQP Benchmark is based on the well-established TPC-H benchmark [24] to streamline its implementation and acceptance. TPC-H is a benchmark portraying the business analytics activity of a wholesale supplier. It consists of 22 business-oriented ad-hoc queries as well as concurrent updates. We propose to use the original TPC-H queries and data generator, but deviate in the benchmark execution to account for the differing goal of measuring elasticity. A number of benchmark proposals take a similar approach for the workload selection (cf. MrBench [13], BigBench [20]).

Following [23], the requirements can vary in one the following three dimensions: (1) The number of concurrent users/query streams, (2) performance requirements (i.e. query execution time) and (3) data volume. Different requirement dimensions lead to different scalability and elasticity patterns. The size of the cluster changes accordingly to compensate for the updated requirement.

3.1 Workload Details

In the following we describe the execution of the workload. For the sake of clarity, we describe a setup to measure the elasticity of the first dimension, i.e. scaling with the number of concurrent users. The workflow can be easily adjusted to test other dimensions by swapping the target requirement.

Loading. First, the data is generated and loaded into the system. We use TPC-H's *DBGen* to generate the initial data set. Subsequently, we execute two consecutive query sets per stream, one to warm up the system and the second to acquire the reference performance at the base level used to determine the scaling for the first phase (see Sect. 3.3).

Queries. The benchmark run consists of a series of phases which in turn are split into a measurement part and a reference part. Each phase starts with the cluster size and client number change by the selected growth factor (cf. Sect. 3.2). The queries are executed until the performance stabilizes. To detect the stable state of the system we monitor query latencies and compare them with a known stable state for the current configuration. The measurement part is followed by the reference part in which we execute the same amount of work and measure the elapsed time as the reference time. In the Sect. 3.3 we describe how the benchmark metrics are calculated from these measurements.

Updates. We simulate the database changes and maintenance by running a parallel update stream during each phase. TPC-H refresh functions one and two and are amended with updates to simulate data cleansing, as the benchmark represents operation on raw, transactional data dumps (Extract-Load-Transform approach). Each stream updates .001 % of the rows during a single phase. Each update operation executes the following: first a random tuple from ORDERS is chosen. The attributes ORDERSTATUS, ORDERDATE and TOTALPRICE are updated with random values. Finally, the attributes QUANTITY, EXTENDEDPRICE and SHIPDATE of the corresponding LINEITEMS-tuple are updated with random values. The updates executed must be visible in the following phase at the latest.

3.2   Parameters

The execution of the EQP Benchmark can be customized by a number of parameters. When reporting the results, the chosen parameters should be exactly specified. Only results obtained using the same parameters can be directly compared. Besides the choice of the requirements dimension (see Sect. 3) the following benchmark parameters can be configured:

1. Amount of data loaded initially (TPC-H scale factor SF)
2. Initial number of worker nodes.
3. Growth factor (GF), i.e. growth rate of number of query streams, worker nodes or data amount between the phases.
4. Number of update streams in each phases.

3.3   Metrics

Our metrics focus on the elasticity of the whole system and the cost of dynamically adding new worker nodes. To evaluate this aspect we determine two metrics: *scaling overhead* and *elastic overhead*. Figure 2 represents a single benchmark phase.

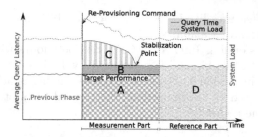

Fig. 2. Metric and measurements in a single phase.eps

Based on the number of nodes, we determine the target query set execution time for each phase, i.e. the execution time of a complete query set on a single stream which we expect from a linearly scaling system after node re-provisioning. With the changing number of query streams or data volume we aim for the times to remain constant. If the goal is query latency reduction, then the target times should drop proportionally with the number of added worker nodes.

The *scaling overhead* is the time wasted while the system stabilizes. For a single phase and a single worker the scaling overhead is the area B in Fig. 2 which we calculate as the difference of the time spent in the measurement part and the target time for the number of queries the system needed to stabilize. The value is then multiplied by the number of active workers in the respective phase. For scaling overhead only the values from the growth phases are considered.

The *elasticity overhead* is the time lost because of the sub-optimal elasticity of the system. It corresponds to the area C. For a single phase and a single worker, the elastic overhead is the difference between the measurement part time and the reference part time. Finally, the sum of the elasticity overheads from every (growth and decline) phase and worker is reported.

Using a specified service price the previously determined time expenditures are converted to the monetary metrics: *scaling costs* and *elasticity costs*. The time and monetary costs can be summarized to *elastic scalability overhead* and *elastic scalability costs* respectively.

4   Evaluation

To evaluate our benchmark design we conducted initial experiments using a private cloud. In the following we present the tested system, preliminary results as well as an evaluation of the results. Note that the results are presented for illustrating the benchmark and should not be used to judge the performance of the tested software.

4.1   System Under Test

Shark [2] is an SQL-like query engine built on top of Apache Spark [9], a general purpose distributed computing cluster. By utilizing Spark's capabilities Shark offers a number of features like in-memory cached tables, mid-query fault tolerance and easy horizontal scalability.

Our cluster runs on four Intel Xeon CPU E5-2660 nodes with 256 GB RAM running Ubuntu 13.10 64-bit and is connected by 1 Gb/s LAN. On the nodes we deployed Shark in version 0.8.0 with the corresponding Spark version (0.8.0). Hadoop 1.2.0 was used as the underlying distributed file system with 4 slaves running on the same nodes.

4.2   Benchmark Execution and Results

For simplicity, we configured no update streams in our benchmark run and limited the query set to TPC-H queries 1 to 5. We made sure that the system was running stable at the end of each measurement phase and no cached results were reused in the following phases, so the tested system did not get any "unfair" advantage.

The benchmark run consists of 3 growth phases and 2 decline phases. The scaling factor is 2, the growth factor is 2 as well. As mentioned in Sect. 3.1, we choose the number of parallel clients as the changing dimension, starting with a single client in the first phase. We made use of the TPC-H port from the Hive developers [12] and extended it with optimizations provided by Shark.

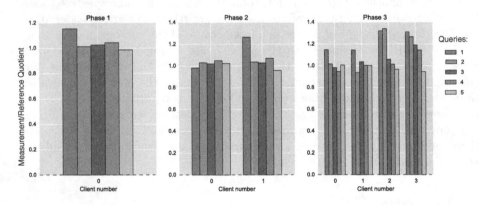

Fig. 3. Measurement-to-reference-time quotients for growth phases.

The results have confirmed our intuition about the elasticity behavior of the tested system. The quotients of the query times of measurement and reference parts are presented in the Fig. 3. The performance of the newly added clients is poor at the beginning and subsequently stabilizes in the course of the execution. Additionally, the elastic overhead is extended by the start-up time of the workers, as well as initial connection times of the new clients. Table 1 shows the total duration of each part and calculated metrics. While the systems scales quite well under the presented conditions, the elasticity overhead adds up to a significant time expenditure.

Table 1. Benchmark execution results (in seconds)

	Phase 1	Phase 2	Phase 3	Phase 4	Phase 5
Number of Nodes	1	2	4	2	1
Measurement time	139.5	136.6	155.16	135.6	125.3
Reference time	125.3	127.4	134.8	129.3	123.3
Elastic Overhead	130.4				
Scaling Overhead	33.8				
Elastic Scaling Overhead	164.2				

5   Conclusions

We propose a new benchmark which specifically aims to measure and evaluate the elasticity of analytical, horizontally scalable systems. We model a typical business case in which the requirements for an analytical query system change and the system is forced to efficiently adapt by scaling in or out. At the same time, we kept the benchmark definition abstract to embrace a larger number of systems and architectures. The resulting metrics are easily understandable and can be directly translated into business terms.

Our initial evaluation shows that even under near-ideal conditions the tested system still have to pay a considerable performance penalty for dynamic horizontal scaling. With the increased volatility of workloads and easy availability of cloud computational resources the elasticity information provided by the EQP Benchmark becomes as crucial as raw performance metrics for a well-founded system choice.

In the future we aim to test a wider variety of systems, including distributed RDBMSs. Additionally, we plan to evaluate public cloud offerings, which are very relevant for a variety of business users with flexible requirements. Finally, the EQP Benchmark should be implemented as a part of an existing benchmark suite (e.g. OLTPBenchmark [1]) to improve re-usability and increase its adoption.

References

1. http://oltpbenchmark.com/. Accessed 7 May 2014
2. Shark. http://shark.cs.berkeley.edu
3. Abouzeid, A., Bajda-Pawlikowski, K.: HadoopDB: an architectural hybrid of MapReduce and DBMS technologies for analytical workloads. In: Proceedings of the VLDB Endowment (2009)
4. Baru, C., Bhandarkar, M., Nambiar, R.: Setting the direction for big data benchmark standards. In: Selected Topics in Performance Evaluation and Benchmarking, pp. 1–13 (2013)
5. Chen, Y., Raab, F., Katz, R.: From tpc-c to big data benchmarks: a functional workload model. In: Rabl, T., Poess, M., Baru, C., Jacobsen, H.-A. (eds.) WBDB 2012. LNCS, vol. 8163, pp. 28–43. Springer, Heidelberg (2014)

6. Cooper, B.F., Silberstein, A., Tam, E., Ramakrishnan, R., Sears, R.: Benchmarking cloud serving systems with YCSB. In: Proceedings of the 1st ACM symposium on Cloud computing - SoCC 2010, p. 143 (2010)
7. Dory, T., Mejías, B., Roy, P.V., Tran, N.: Measuring elasticity for cloud databases. In: CLOUD COMPUTING 2011 : The Second International Conference on Cloud Computing, GRIDs, and Virtualization, pp. 154–160 (2011)
8. Floratou, A., Teletia, N., DeWitt, D.: Can the elephants handle the NoSQL onslaught? Proc. VLDB Endow. **5**(12), 1712–1723 (2012)
9. Foundation, A.: Spark: Lightning-fast cluster computing (2014). http://spark.apache.org/. Accessed 21 March 2014
10. Gambi, A., Moldovan, D., Copil, G., Truong, H.-L., Dustdar, S.: On estimating actuation delays in elastic computing systems. In: 8th International Symposium on Software Engineering for Adaptive and Self-Managing Systems, pp. 33–42 (2013)
11. Islam, S., Lee, K., Fekete, A., Liu, A.: How a consumer can measure elasticity for cloud platforms. In: Proceedings of the Third Joint WOSP/SIPEW International Conference on Performance Engineering - ICPE 2012, p. 85 (2012)
12. Jia, Y.: Running the TPC-H Benchmark on Hive. Corresponding issue (2009). https://issues.apache.org/jira/browse/HIVE-600
13. Kim, K., Jeon, K., Han, H., Kim, S.-G.: Mrbench: a benchmark for mapreduce framework. In: Proceedings of the 2008 14th IEEE International Conference on Parallel and Distributed Systems, ICPADS 2008, pp. 11–18 (2008)
14. Konstantinou, I., Angelou, E.: On the elasticity of NoSQL databases over cloud management platforms. In: Proceedings of the 20th ACM international conference on Information and Knowledge Management, pp. 2385–2388 (2011)
15. Meisner, D., Sadler, C.M., Barroso, L.A., Weber, W.-D., Wenisch, T.F.: Power management of online data-intensive services. In: Proceeding of the 38th Annual International Symposium on Computer Architecture - ISCA 2011, p. 319 (2011)
16. Mühlbauer, T., Rödiger, W., Reiser, A.: ScyPer: elastic OLAP throughput on transactional data. In: Proceedings of the Second Workshop on Data Analytics in the Cloud, pp. 1–5 (2013)
17. O'Neil, P., O'Neil, E., Chen, X., Revilak, S.: The star schema benchmark and augmented fact table indexing. In: Nambiar, R., Poess, M. (eds.) TPCTC 2009. LNCS, vol. 5895, pp. 237–252. Springer, Heidelberg (2009)
18. Ousterhout, J.K., Agrawal, P., Erickson, D., Kozyrakis, C., Leverich, J., Mazières, D., Mitra, S., Narayanan, A., Parulkar, G.M., Rosenblum, M., Rumble, S.M., Stratmann, E., Stutsman, R.: The case for ramclouds: scalable high-performance storage entirely in dram. Operating Syst. Rev. **43**(4), 92–105 (2009)
19. Pavlo, A., Paulson, E., Rasin, A., Abadi, D.J., DeWitt, D.J., Madden, S., Stonebraker, M.: A comparison of approaches to large-scale data analysis. In: Proceedings of the 35th SIGMOD international conference on Management of data, p. 165 (2009)
20. Rabl, T., Ghazal, A., Hu, M., Crolotte, A.: Bigbench specification V0. 1. In: Specifying Big Data Benchmarks (2012)
21. Rödiger, W., Mühlbauer, T., Unterbrunner, P.: Locality-sensitive operators for parallel main-memory database clusters (2014)
22. Stonebraker, M.: Mapreduce and parallel dbmss: friends or foes? Commun. ACM **53**(4), 10 (2010)
23. Tinnefeld, C., Kossmann, D., Grund, M., Boese, J.-H., Renkes, F., Sikka, V., Plattner, H.: Elastic online analytical processing on ramcloud. In: Guerrini, G., Paton, N.W. (eds.), EDBT, pp. 454–464. ACM (2013)
24. Transaction Processing Performance Council. TPC-H specification (2010). www.tpc.org/tpch

An Approach to Benchmarking Industrial Big Data Applications

Umeshwar Dayal$^{(\boxtimes)}$, Chetan Gupta, Ravigopal Vennelakanti,
Marcos R. Vieira, and Song Wang

Big Data Research Lab, Hitachi America, Ltd., R&D, Santa Clara, CA, USA
umeshwar.dayal@hal.hitachi.com,
{chetan.gupta,ravigopal.vennelakanti,
marcos.vieira,song.wang}@hds.com

Abstract. Through the increasing use of interconnected sensors, instru-
mentation, and smart machines, and the proliferation of social media and
other open data, industrial operations and physical systems are generating ever increasing volumes of data of many different types. At the
same time, advances in computing, storage, communication, and big
data technologies are making it possible to collect, store, process, analyze and visualize enormous volumes of data at scale and at speed. The
convergence of Operations Technology (OT) and Information Technology (IT), powered by innovative data analytics, holds the promise of
using insights derived from these rich types of data to better manage our
systems, resources, environment, health, social infrastructure, and industrial operations. Opportunities to apply innovative analytics abound in
many industries (e.g., manufacturing, power distribution, oil and gas
exploration and production, telecommunication, healthcare, agriculture,
mining) and similarly in government (e.g., homeland security, smart
cities, public transportation, accountable care). In developing several
such applications over the years, we have come to realize that existing benchmarks for decision support, streaming data, event processing,
or distributed processing are not adequate for industrial big data applications. One primary reason being that these benchmarks individually
address narrow range of data and analytics processing needs of industrial big data applications. In this paper, we outline an approach we are
taking to defining a benchmark that is motivated by typical industrial
operations scenarios. We describe the main issues we are considering
for the benchmark, including the typical data and processing requirements; representative queries and analytics operations over streaming
and stored, structured and unstructured data; and the proposed simulator data architecture.

1 Introduction

Today, we are at the dawn of transformative changes across industries, from
agriculture to manufacturing, from mining to energy production, from healthcare to transportation. These transformations hold the promise of making
our economic production more efficient, cost effective, and, most importantly,

T. Rabl et al. (Eds.): WBDB 2014, LNCS 8991, pp. 45–60, 2015.
DOI: 10.1007/978-3-319-20233-4_6

sustainable. These transformations are being driven by the convergence of the global industrial system (Operations Technology (OT)) with the power of integrating advanced computing, analytics, low-cost sensing and new levels of connectivity (Information Technology (IT)).

Through the increasing use of interconnected sensors and smart machines and the proliferation of social media and other open data, industrial operations and physical systems produce a very large volume of continuous stream of sensor, event and contextual data. This unprecedented amount of rich data needs to be stored, managed, analyzed and acted upon for sustainable operations of these systems. Big data technologies, driven by innovative analytics, are the key to creating novel solutions for these systems that achieve better outcomes at lower cost, substantial savings in fuel and energy, and better performing and longer-lived physical assets.

Opportunities to create big data solutions abound in many industries (e.g., power distribution, oil and gas exploration and production, telecommunication, healthcare, agriculture, mining) and in the public sector (e.g., homeland security, smart cities, public transportation, population health management). To realize operational efficiencies and to create new revenue-generating lines of business from the deluge of data requires the convergence of IT $\times$ OT. The convergence of these two technologies can be obtained by leveraging an analytics framework to translate data-driven insights from a multitude of sources into actionable insights delivered at the speed of the business. Thus, innovations in analytics will be required: (1) to deal with the vast volumes, variety, and velocity of data; and (2) to create increasing value by moving from descriptive or historical analytics (e.g., what has happened and why?) to predictive analytics (e.g., what is likely to happen and when?) and finally to prescriptive analytics (e.g., what is best course of action to take next?).

In this paper, we provide a detailed discussion of a proposed benchmark for industrial big data applications. Existing benchmark proposals either focus on OLTP/OLAP workloads for database systems or focus on enterprise big data systems. Our objective is to define a benchmark for IT $\times$ OT big data applications in industrial systems. We recognize that proposing such a benchmark is a complex and evolving task. To the best of our knowledge, this paper is the first to outline a systematic approach to defining a benchmark for industrial big data applications.

The paper is organized as follows. Section 2 details existing benchmarking proposals for various application domains. Section 3 describes the characteristics of IT $\times$ OT Big Data Applications. Section 4 provides an overview of the main features of our proposed benchmark. Section 5 details the initial benchmark architecture implementation. Finally, Sect. 6 concludes the paper and outlines the future extensions to our proposed benchmark.

2   Related Work

There exist many efforts for developing benchmarks for big data systems, each focusing on evaluating different features. Examples of industry standard benchmarks are TPC-H [13,18] and TPC-DS [10,14], both developed by the Transaction Processing Performance Council (TPC), the leading benchmarking council for transaction processing and database benchmarks. These two decision support benchmarks are employed to compare SQL-based query processing performance in SQL-on-Hadoop and relational systems [1,6,12,15,17]. Although these benchmarks have been often used for comparing query performance of big data systems, they are basically SQL-based benchmarks and, thus, lack the new characteristics of industrial big data systems.

Several proposals have extended the TPC-H and TPC-DS to deal with new characteristics of big data systems. For example, BigDS benchmark [20] extends TPC-DS for applications in the social marketing and advertisement domains. Han and Lu [8] discuss key requirements in developing benchmarks for big data systems. However, neither of these two proposals defines a query set and data model for the benchmark. HiBench [9] and BigBench [2,4,7] are *end-to-end application-based benchmarks* that extend the TPC-DS model to generate and analyze web logs. Some other works propose *component benchmarks* to analyze specific features or hardware of big data systems (e.g., [12,19]). However, a *component benchmark*, which measures performance of one (or a few) components, has limited scope compared to *end-to-end benchmarks*, which measures full system performance. Additionally, there are proposals of big data tools for benchmarking big data systems (e.g., [16,21]), and even efforts to maintain a top-ranked list of big data systems based on a benchmark [3] (similar in concept to the Sort Benchmark competition [11]).

Recently the TPC released the TPC Express Benchmark for Hadoop Systems (TPCx-HS) [5]. This benchmark is based on the TeraSort benchmark, which is a Hadoop-based sort benchmark [11], to measure hardware, operating system and commercial Hadoop File System API.

Common to all of the above benchmarks is that they are designed to measure limited features of big data systems of specific application domains (e.g., decision support, streaming data, event processing, distributed processing). However, none of existing benchmarks covers the range of data and analytics processing characteristic of industrial big data applications. To the best of our knowledge, our proposed benchmark is the first to address typical data and processing requirements, representative queries and analytics operations over streaming and stored, structured and unstructured data of industrial big data applications.

3 IT × OT Big Data Applications

In this section, we first describe the main characteristics of industrial big data systems in terms of: (1) latencies and kinds of decisions; (2) variety of data; and (3) type of operations that occurs in these systems. We then illustrate examples of these characteristics using two industrial applications.

The class of IT × OT big data applications can be understood in terms of the life cycle of data, as illustrated in Fig. 1. In today's architecture, the outermost cycle is the strategy cycle, where the senior executives and back-office staff use

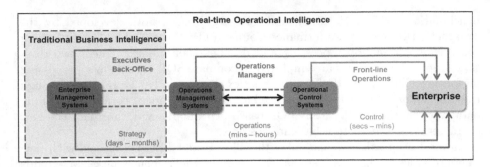

Fig. 1. For IT×OT applications "Operational Intelligence" is the key requirement. When decreasing time scales for responses (decision latencies), we have an increase in automation for big data systems.

historical data stored in an Enterprise Data Warehouse (EDW) to take long term decisions (i.e., very long latency – days or months). In this class of applications the data is stored in an EDW which has been already processed, normalized and structured using ETL tools. Furthermore, the operations available in these systems are limited to historical data analysis.

We often see application needs moving from traditional business intelligence to a more real-time operational intelligence. Thus, we want to be able to make decisions in order of minutes or hours for **operations management systems**, and if we want to decrease decision latencies, in the order of seconds or minutes, we have **operational control systems**. Along with, decreasing the latencies and kinds of decisions when moving to a more real-time operational intelligence, we also observe an increase in variety of data (e.g., stored/streaming data, structured/semi-structured/unstructured data) and in the complexity of operations (e.g., ranging from advanced queries to time series analysis to event correlation and prediction) in these systems. These advanced features in real-time operational intelligence are illustrated in the middle and innermost cycles in Fig. 1. In the middle cycle, or daily operations cycle, the operations managers take day-to-day decisions, such as inventory management, whereas in the innermost cycle, the control or the operations management cycle, which is aimed at responding in real time to changes in the state of the enterprise for efficient management of resources.

The traditional business intelligence approach to building applications has the main drawback that EDWs only store part of the data, while optimal decision making at any time scale is dependent on all the data available to the enterprise. Hence, the new architecture for IT × OT application can exploit the emerging big data architectures to make available all the data collected from the enterprise and perform analytics over this data for decision making. In other words, we are moving from traditional business intelligence towards real-time operational intelligence, where a single system is able to provide decision making at any time scale. We illustrate this with examples from two industries. For both of these industries we illustrate the data and problem requirement that cannot be address by today's enterprise architectures.

3.1   Electric Power Industry Application

Applications in the electric power industry are collecting data at fine granularities, from a large number of smart meters, power plants, and transmission and distribution networks. The variety of data collected also ranges from streaming data (e.g., data from a large volume of smart meters and sensors, energy trading time series) to stored data (e.g., daily billing information). Moreover, applications in this domain are increasing in complexity, from simple historical queries to more advanced analytics on streaming data (e.g., list the neighborhoods with the highest increase energy demands compared to the same period last week), to time series and event predictions (e.g., when and where a power failure will occur?).

In terms of a system for operational intelligence, in the outermost cycle (illustrated in Fig. 2) on a long term basis we are concerned with problems such as matching power supply with demand for a region, allocating new suppliers, and managing future derivative contracts. In the daily cycle, we are concerned with daily billing and managing daily supply with demand. In the operation management cycle, we address problems such as detection of outages, dynamic pricing, consumption optimization, personalized electric power usage recommendations, among others.

Today, in the electricity industry, there are different systems for each of these granularities of decision making. Furthermore, the raw smart meter data is stored only at aggregate level, and the original raw data is discarded, consequently not available for decision making.

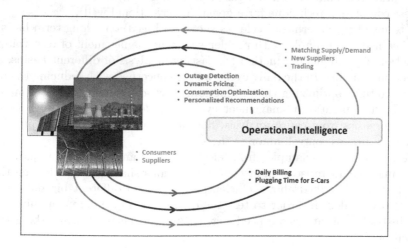

Fig. 2. Smart power grid application scenario.

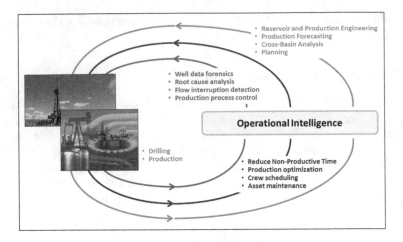

Fig. 3. Smart oil and gas fields application scenario.

3.2   Oil and Gas Fields Application

Applications in the oil and gas industry are collecting data at fine granularities, from electric submersible pumps, oil and gas pipelines, sensors in well reservoirs, drilling rigs, oil platforms, and oil and gas refineries. The variety of data being collected is also changing from only historical to streaming data (e.g., data from multiple sensors installed in a network of oil rigs), and from only structured to also include unstructured data (e.g., analysis on data from scientific papers, operator notes, technical reports). Also, the complexity on the type data operations are increasing, from queries on historical data to more data-driven intensive computation (e.g., find the oil and gas plays with high hydrocarbon content, find the best completion technique for a given unconventional well).

In terms of the outermost cycle (as shown in Fig. 3) on a long term basis we are concerned, for example, with problems such as management of reservoir and production, oil and gas production forecasting, analysis of different basins, and production planning. In the daily cycle, we are concerned with reducing the non-productive time, optimizing the production, crew scheduling, and asset maintenance. In the operation management cycle, we address problems such as well as data forensics, root-cause analysis, flow interruption detection in pipelines, production process control, among others.

We saw with the examples from electric power and oil and gas industries the common characteristics industrial applications share. To handle the three characteristics described above, efficient, scalable and different big data architectures are needed. In order to test these complex and different architectures for industrial applications we need a generic approach to benchmarking these applications.

4   Features of Industrial Big Data Benchmark

We now present the required features needed to build an industrial big data benchmark. We first detail an **application scenario** to illustrate the different requirements of the IT × OT applications. We then abstract the application scenario and develop an **activity model**, which gives a definition of the important features and their relationships in the application scenario. We then discuss the **streaming and historical data** that is potentially collected for such applications, and the **queries and analytics operations** that can be performed over the data. Finally, we discuss the **evaluation measures** that should be used to evaluate the benchmark and the **simulator parameters** that need to be set for the simulator and data generator in order to obtain the benchmark with the desired properties.

4.1   Application Scenario

We select an application scenario from the surface mining industry because it is representative of many industrial IT × OT systems: it is characterized by a set of field operations involving physical objects, such as stationary and moving equipment; the operational activities generate data of diverse types, and require a variety of analytics processing to provide operational intelligence at different latencies, both in the field and at remote operations centers.

Surface mining is a category of mining in which soil overlying valuable natural resources is removed. One process commonly employed in the surface mining industry is known as *open-pit mining*. In open-pit mining, rocks are first blasted where the raw material is to be found. Once the blast has taken place, the regular mine operation happens in shifts. During a typical shift several shovels or loaders pick up either the raw material or the waste from the blast site and load it into trucks or haulers. If the truck is hauling raw material it proceeds to the processing plant or the stockpile; if the truck is hauling waste it proceeds to a dump site. Once the truck is emptied, it goes back to the loading unit (shovel or loader) and the cycle (referred to as an operation cycle) starts again.

At any point in the mine, there are several shovels and a much larger number of trucks operating. In a modern mine, data is constantly being collected from various different equipment. From a mine managers' perspective, there are several primary objectives that need to be optimized during the operation of the mine: (1) getting the correct mix of different grades of the raw material based on the market conditions; (2) making sure the equipment is available in the field to perform the activities of operation processes; and (3) making the most efficient use of the equipment that is available in the mine.

An abstract view of data collection and processing for such field application scenario is illustrated in Fig. 4. Data from various different field equipment is constantly relayed via field sensors to a Remote Operations Center (ROC). Various decisions support applications are built on top of the collected data.

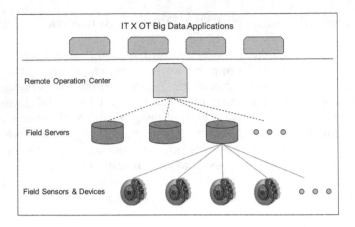

Fig. 4. Generic field IT × OT big data application scenario.

4.2 Activity Model

From the described application scenario we can highlight some of the high level requirements for an IT × OT application:

1. Data ingestion and storage capabilities from several different processes and equipment;
2. Data processing features from different data types and sources;
3. Encoding some domain related knowledge regarding the operations and activity on the field site;
4. Data analysis from simple queries to advanced analytics with different execution and response types;
5. Some simple to advanced visualization features from the data operations and activities[1].

From the above list of requirements, we further abstract and define an activity model:

- **Operational process** is a Directed Acyclic Graph (DAG) of activities, where manufacturing operations are the vertices and the edges represent a process sequence. In the mining use case, an operational process for loaders and haulers, see Fig. 5, has the following graph representation: *loading → hauling full → queuing at dump → dumping → hauling empty → queuing at loader*;
- Each **activity** is performed by **equipment**, which acts on an **object**. In the mining use case, **equipment** is a shovel or a hauler, and **object** is waster or raw materials;
- **Equipment** might be in **stationary** or **moving** states. Shovels are often in **stationary** state and haulers are in **moving** state;
- **Sensors** measure attributes of **equipment** and attributes of **object**;

---

[1] Visualization features are not covered in this version of our proposed benchmark.

Fig. 5. Operational process of a haul cycle represented as a DAG of activities.

- **Events** signal start and finish of **activities**;
- **Equipment cycle** is one complete sequence of **activity** for **equipment**;
- **Operational cycle** is one complete sequence of equipment **activities** for an **operational process**;
- **Work shift** consists of several different **operational cycles** running concurrently and repeatedly.

Based on the activity model, we further detail other characteristics of the proposed benchmark for IT × OT applications.

4.3   Streaming and Historical Data

There are several different types of data that are generated by OT systems. Our proposed benchmark uses data generated by a simulator that simulates the operation processes (described in Sect. 5). Historical data is stored in the data management and processing layers (e.g., RDBMS, Key-Value Store, Hadoop) and real time data is streamed. In the following we describe the data schema, shown in Fig. 6, for the historical and streaming data:

1. Operational data from equipment includes both real-time stream data (e.g., from current work shift) and historical, stored data (e.g., from n previous work shifts):
   - Event data generated from equipment (begin/end activities, other operational events):
     ActivityEventMeasure (equipmentId, eventId, beginTimestamp, endTimestamp)
   - Measurements from sensors on equipment:
     EquipmentSensorMeasure (sensorId, value, timestamp)
   - GPS Sensors on moving equipment:
     MovingSensorLocationMeasure (sensorId, GPSCoordinates, timestamp)
   - Field sensors (e.g., beacons):
     FieldEquipmentMeasure (sensorId, equipmentId, timestamp)

- Equipment data:
  EquipmentFleetMeasure (equipmentId, fleetId, timestamp)
2. Non-equipment operational data:
   - Measurements from sensors that measure attributes of object being acted upon:
     ObjectAttributeEstimationMeasure (sensorId, value, timestamp)
3. Object operational data:
   - Production target within a shift:
     ShiftProductionTargetMeasure (shiftId, targetProduction)
   - Relationship between activity and event type:
     EventActivityMapping (acitivityId, eventId)
4. External operational data:
   - Ambient sensor measures:
     AmbientSensorMeasure (sensorId, value, timestamp)
5. Process operational data:
   - Shift start and end times:
     Shift (shiftId, beginTimestamp, endTimestamp)
6. Operator data:
   - Operational notes:
     OperationalNotes (authorId, timestamp, noteValue)
7. Business data:
   - Attributes of equipment:
     Equipment (equipmentId, type, make, year, properties)
   - Sensor attributes:
     Sensor (sensorId, measurementType, make, model, properties)
     EquipmentSensorRelationship (sensorId, equipmentId, timestamp)
     FieldSensorLocation (sensorId, GPSLocation, timestamp)
   - Relationship between different equipment:
     EquipmentAdjacency (equipmentId1, equipmentId2, activityId)
   - Process model:
     ActivityGraph (activityId1, activityId2, adjacencyRelationship)
8. There may exist other data sources:
   - External business data (e.g., equipment cost):
     CostOfEquipment (equipmentId, cost, timestamp)
   - Manufacturer's manual:
     EquipmentUserManual (equipmentId, partId, description)

4.4   Queries and Analytics Operations

Having defined the data schema for IT × OT benchmark, we now describe the operations on the data, ranging from simple SQL-based queries to more complex analytics operations. Even for a single type of operation (e.g., a single SQL-based query), the possible range of queries and analytics operations form a large space. To systematically study this, we describe the queries and analytics operations in terms of the following properties:

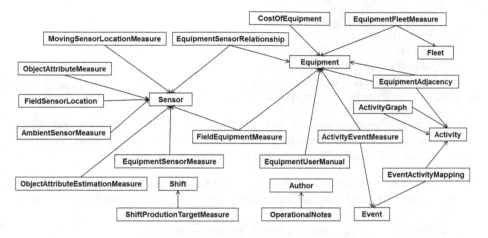

Fig. 6. Logical data schema.

1. Data types:
   - Sensor data
   - Event data
   - Log data
   - Geospatial data
   - Business data
2. Data states:
   - Data in motion
   - Data at rest
   - Hybrid data state
3. Query execution types:
   - One-shot query
   - Continuous query
   - Scheduled query
4. Query response types:
   - Real-time query
   - Near real-time query
   - Best effort query

We identified some important types of queries and analytics operations that are executed over the stored and streaming data generated by the simulator. Below we summarize the types of queries and analytics operations:

1. Aggregate functions;
2. Lookup & search operations;
3. ETL operations on new data;
4. OLAP on stored data;
5. Data mining and machine learning operations:
   - Outlier detection;

- Time series prediction;
- Event prediction;
- Key Performance Indicator (KPI) prediction;
- Spatio-temporal pattern detection;
- Diagnostics and root cause analysis.
6. Planning and optimization operations.

We now present several sample benchmark queries and analytics operations. These sample operations are defined in terms of the query description and primary data entity accessed to answer such queries:

1. Aggregate function queries performed over time: last $t = 5\,\mathrm{min}$ (sliding window move by 1 min), current shift (landmark window), or previous shift (tumbling window), compute:
   (a) Total number of activities by each equipment:
       (ActivityEventMeasure, EventActivityMapping)
   (b) Total number of operation cycles:
       (ActivityEventMeasure, ActivityAdjacency, EventActivityMapping, ActivityGraph)
   (c) Total number of equipment cycles per equipment per equipment type:
       (ActivityEventMeasure, Equipment, EventActivityMapping, Activity-Graph)
   (d) Total distance moved per each moving equipment per operation cycle:
       (ActivityEventMeasure, MovingSensorLocationMeasure, EquipmentSensorRelationship)
   (e) Top-k average high measure from equipment:
       (EquipmentSensorMeasure, EquipmentSensorRelationship)
2. Detection of threshold events:
   (a) Alert if the object attribute measure is outside the bounds (mean $+/- 2\sigma$ over a fixed window of last shift):
       (ObjectAttributeEstimationMeasure)
   (b) Alert if wait time for any activity is above its expected value bounds (mean $+/- 2\sigma$ over a sliding window of last k hour sliding by 1 h):
       (ActivityEventMeasure)
3 Detection of outlier events:
   (a) Real-time outlier detection for machines based on equipment sensor health data:
       (EquipmentSensorMeasure, EquipmentSensorRelationship)
4. Detection of general events:
   (a) Alert if there is an order violation in terms of activity order:
       (ActivityEventMeasure, ActivityGraph, EventActivityMapping)
   (b) Alert if moving equipment visits a specified field sensor less than once during an operational cycle:
       (FieldEquipmentMeasure, ActivityEventMeasure)
5. Production estimation query:
   (a) Predict if an object attribute measure will miss the shift target:
       (ObjectAttributeEstimationMeasure, ShiftProductionTargetMeasure)

6. Facility optimization query:
   (a) Maximize the utilization of the most expensive equipment in the working shift:
   (Equipment, CostOfEquipment, ActivityData)
7. Sensor health query:
   (a) Detect faulty sensors by comparing to similar sensors:
   (EquipmentSensorMeasure, EquipmentSensorRelationship, Sensor)
8. Equipment event prediction:
   (a) Predict an equipment sensor health event with a time horizon of t hours:
   (EquipmentSensorMeasure, ActivityEventMeasure, EquipmentAttribute, EquipmentSensorRelationship, AmbientSensorMeasure)
9. Recommended action operation:
   (a) For a predicted equipment maintenance event (e.g., downtime), recommend best action:
   (OperatorNotes, EquipmentSensorMeasure, ActivityEventMeasure, EquipmentAttribute, EquipmentSensorRelationship, AmbientSensorMeasure)

4.5   Evaluation Measures

Below we propose evaluation measures for each query and analytics operations previous described:

1. For aggregate functions over time windows (query type 1):
   - **Response time**: report the result within t time units of close of window;
   - **Accuracy**: results should be correct for every query.
2. For threshold, outlier and general event detection (queries type 2–4):
   - **Response time**: report the event within t time units of event occurring;
   - **Accuracy**:
     • All event occurrences should be detected;
     • Report f-measure for outlier events.
3. For production estimation query (query type 5):
   - **Response time**: report the production estimation within t time units of work shift;
   - **Accuracy**: report f-measure for estimation.
4. For facility optimization query (query type 6):
   - **Response time**: report the utilization value within t time units of work shift;
   - **Relative error**: report the difference between the reference and reported maximum utilization value.
5. For sensor health query (query type 7):
   - **Response time**: report the sensor health within t time units of close of window;
   - **Accuracy**: results should be correct for every query.

6. For equipment event prediction (query type 8):
   - **Response time**: report the prediction within t time units of close of window;
   - **Accuracy**: report f-measure for prediction.
7. For recommended action operation (query type 9):
   - **Response time**: report the utilization value within t time units of work shift;
   - **Accuracy**: report f-measure for recommended action.

4.6   Simulation Parameters

As part of the proposed benchmark we provide a simulator and data generator. The activity simulator will simulate the operational processes (e.g., haul cycles) and from that generate the relevant data (e.g., sensor, event, log data). The simulator and data generator is capable of producing large amounts of data in a scalable and high performance fashion. The input parameters are described below:

1. Number of static and moving equipment in the field;
2. Number of sensors for each equipment;
3. Number of field sensors;
4. Sampling frequency of each sensor;
5. Event rates of non-activity events;
6. Number of activities in an equipment cycle;
7. Number of activities in an operational cycle;
8. Number of operational cycles in a shift;
9. Number of shifts in the entire benchmark;
10. Volume of historical and business data.

5   Benchmark Implementation

In our proposed benchmark architecture, shown in Fig. 7, we provide two modules as part of the benchmark: the simulator and data generator and the evaluation modules. The simulator and data generator module receive a set of input

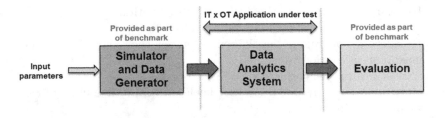

Fig. 7. Benchmarking IT × OT big data applications: simulator and data generator and evaluation modules are provided as part of the benchmark.

parameters and output a set of (streaming/historical) logs, events, business data. The set of output data includes different data types, from static business data to real-time stream and event data. This set of data is provided as input to the IT × OT application under test. The second module evaluates and generates reports on the output data produced by the IT × OT application.

6   Conclusion and Future Works

In this paper, we present the first steps towards the definition of a benchmark for industrial big data applications. Previous benchmarks do not cover the wide range of data, queries and analytics processing characteristics of industrial big data applications. Our proposed benchmark is motivated by a typical industrial operations scenario, including the typical data types and processing requirements, representative queries and analytics operations, and a data generator and simulator architecture.

As a next step in this research we are performing the following tasks:

1. We have provided a description of a benchmark. Based on the community's feedback we will refine the benchmark and provide precise definition of the set of queries and analytical operations;
2. We are starting to implement the benchmark. As part of the implementation, we are developing the data generator and simulator modules;
3. We are studying additional benchmarking measures targeting industrial applications (e.g., ease of development, cost, energy efficiency, security, maintainability).

References

1. Abouzied, A., Bajda-Pawlikowski, K., Huang, J., Abadi, D.J., Silberschatz, A.: HadoopDB in action: building real world applications. In: Proceedings of the ACM SIGMOD International Conference on Management of Data, pp. 1111–1114 (2010)
2. Baru, C., et al.: Discussion of bigbench: a proposed industry standard performance benchmark for big data. In: Nambiar, R., Poess, M. (eds.) TPCTC 2014. LNCS, pp. 44–63. Springer, Hiedelbeg (2015)
3. Baru, C., Bhandarkar, M., Nambiar, R., Poess, M., Rabl, T.: Big data benchmarking and the BigData top100 list. Big Data J. 1(1), 60–64 (2013)
4. Chowdhury, B., Rabl, T., Saadatpanah, P., Du, J., Jacobsen, H.A.: A BigBench implementation in the Hadoop ecosystem. In: Rabl, T., Raghunath, N., Poess, M., Bhandarkar, M., Jacobsen, H.A., Baru, C. (eds.) WBDB 2013. LNCS, vol. 8585, pp. 3–18. Springer, Heidelberg (2014)
5. Transaction Processing Performance Council, TPCx-HS, February 2015. www.tpc.org/tpcx-hs/
6. Floratou, A., Minhas, U.F., Özcan, F.: SQL-on-hadoop: full circle back to shared-nothing database architectures. Proc. VLDB Endowment (PVLDB) 7(12), 295–1306 (2014)

7. Ghazal, A., Rabl, T., Hu, M., Raab, F., Poess, M., Crolotte, A., Jacobsen, H.-A. BigBench: towards an industry standard benchmark for big data analytics. In: Proceedings of the ACM SIGMOD International Conference on Management of Data, pp. 1197–1208 (2013)
8. Han, R., Lu, X., Xu, J.: On big data benchmarking. In: Han, R., Lu, X., Xu, J. (eds.) BPOE 2014. LNCS, pp. 3–18. Springer, Heidelberg (2014)
9. Huang, S., Huang, J., Dai, J., Xie, T., Huang, B.: The HiBench benchmark suite: characterization of the MapReduce-based data analysis. In: Proceedings of the IEEE International Conference on Data Engineering (ICDE) Workshop, pp. 41–51 (2010)
10. Nambiar, R.O., Poess, M.: The making of TPC-DS. In: Proceedings of the International Conference on Very Large Data Bases (VLDB), pp. 1049–1058 (2006)
11. Nyberg, C., Shah, M., Govindaraju, N.: Sort Benchmark, February 2015. http://sortbenchmark.org
12. Pavlo, A., Paulson, E., Rasin, A., Abadi, D.J., DeWitt, D.J., Madden, S., Stonebraker, M.: A comparison of approaches to large-scale data analysis. In: Proceedings of the of the ACM SIGMOD International Conference on Management of Data, pp. 165–178 (2009)
13. Poess, M., Floyd, C.: New TPC benchmarks for decision support and web commerce. SIGMOD Rec. **29**(4), 64–71 (2000)
14. Poess, M., Nambiar, R.O., Walrath, D.: Why you should run TPC-DS: a workload analysis. In: Proceedings of the VLDB Endowment (PVLDB), pp. 1138–1149 (2007)
15. Poess, M., Rabl, T., Caufield, B.: TPC-DI: the first industry benchmark for data integration. Proc. VLDB Endowment (PVLDB) **7**(13), 1367–1378 (2014)
16. Rabl, T., Jacobsen, H.-A.: Big data generation. In: Rabl, T., Poess, M., Baru, C., Jacobsen, H.-A. (eds.) WBDB 2012. LNCS, vol. 8163, pp. 20–27. Springer, Heidelberg (2014)
17. Thusoo, A., Sarma, J.S., Jain, N., Shao, N., Chakka, P., Zhang, N., Antony, S., Liu, H., Murthy, R.: Hive - a petabyte scale data warehouse using hadoop. In: Proceedings of the IEEE International Conference on Data Engineering (ICDE), pp. 996–1005 (2010)
18. Transaction Processing Performance Council (TPC), TPC-H benchmark specification (2008). http://www.tpc.org/tpch/
19. Wang, L., Zhan, J., Luo, C., Zhu, Y., Yang, Q., He, Y., Gao, W., Jia, Z., Shi, Y., Zhang, S., Zheng, C., Lu, G., Zhan, K., Li, X., Qiu, B.: BigDataBench: a big data benchmark suite from internet services. In: IEEE International Symposium on High Performance Computer Architecture (HPCA), pp. 488–499, February 2014
20. Zhao, J.-M., Wang, W.-S., Liu, X., Chen, Y.-F.: Big data benchmark - Big DS. In: Rabl, T., Raghunath, N., Poess, M., Bhandarkar, M., Jacobsen, H.-A., Baru, C. (eds.) WBDB 2013. LNCS, vol. 8585, pp. 49–57. Springer, Switzerland (2014)
21. Zhu, Y., Zhan, J., Weng, C., Nambiar, R., Zhang, J., Chen, X., Wang, L.: BigOP: generating comprehensive BigData workloads as a benchmarking framework. In: Bhowmick, S.S., Dyreson, C.E., Jensen, C.S., Lee, M.L., Muliantara, A., Thalheim, B. (eds.) DASFAA 2014, Part II. LNCS, vol. 8422, pp. 483–492. Springer, Heidelberg (2014)

Hadoop and MapReduce

Benchmarking SQL-on-Hadoop Systems: TPC or Not TPC?

Avrilia Floratou[1]([✉]), Fatma Özcan[1], and Berni Schiefer[2]

[1] IBM Almaden Research Center, 650 Harry Road, San Jose, CA, USA
{aflorat,fozcan}@us.ibm.com
[2] IBM Toronto Lab, 8200 Warden Ave, Markham, ON, Canada
schiefer@ca.ibm.com

Abstract. Benchmarks are important tools to evaluate systems, as long as their results are transparent, reproducible and they are conducted with due diligence. Today, many SQL-on-Hadoop vendors use the data generators and the queries of existing TPC benchmarks, but fail to adhere to the rules, producing results that are not transparent. As the SQL-on-Hadoop movement continues to gain more traction, it is important to bring some order to this "wild west" of benchmarking. First, new rules and policies should be defined to satisfy the demands of the new generation SQL systems. The new benchmark evaluation schemes should be inexpensive, effective and open enough to embrace the variety of SQL-on-Hadoop systems and their corresponding vendors. Second, adhering to the new standards requires industry commitment and collaboration. In this paper, we discuss the problems we observe in the current practices of benchmarking, and present our proposal for bringing standardization in the SQL-on-Hadoop space.

Keywords: SQL · Hadoop · Benchmark · TPC · SPEC · STAC · TPC-DS · TPC-H

1 Introduction

Benchmarks are an integral part of software and systems development, as they provide a means with which to evaluate systems performance in an objective way. While the discussion and the work on new big data benchmarks are in progress, many vendors use the Transaction Processing Performance Council (TPC) [22] benchmark schema, data generator, and queries but are very selective about which parts of the specification and disclosure rules they follow. The TPC was formed to help bring order and governance on how performance testing should be done and results published. Without the rules, the results are not comparable, and not even meaningful.

In the relational database world there was a transition from the "lawless" world of Debit-Credit to the much more rigorous and unambiguous world of TPC-A/TPC-B [11].

Central to the concepts pioneered by the TPC include:

© Springer International Publishing Switzerland 2015
T. Rabl et al. (Eds.): WBDB 2014, LNCS 8991, pp. 63–72, 2015.
DOI: 10.1007/978-3-319-20233-4_7

1. The notion of a specification that is sufficiently high level to permit multiple vendor participation while simultaneously ensuring a high degree of comparability.
2. The concept of "full disclosure" or disseminating sufficient detail that it should be possible to both understand and potentially duplicate the published results.
3. The requirement to "audit" results to ensure adherence to the specification.

In addition, the TPC Policies describe the manner in which the benchmark results can and can not be compared in a public forum. These "Fair Use" rules set the standard for what is and is not allowed[1]. In particular there is a requirement for:

1. **Fidelity:** Adherence to facts; accuracy
2. **Candor:** Above-boardness; needful completeness
3. **Due Diligence:** Care for integrity of results
4. **Legibility:** Readability and clarity

In contrast to this well-regulated relational database benchmarking environment, in the world of SQL-on-Hadoop we are in the "wild west". Example of this is the (mis)use of the TPC benchmarks by the SQL-on-Hadoop systems. Some example SQL-on-Hadoop systems include IBM Big SQL [3,12,15], Hortonworks Hive [2],Cloudera Impala [5], Presto [18], Microsoft Polybase [8], and Pivotal HAWQ [17].

The relatively "free" access and EULA (End User License Agreements) rules of newer systems that are not constrained by the famous "DeWitt" clause[2] makes it easy for the SQL-on-Hadoop vendors to conduct performance comparisons between their system and competitors and publish the results. In fact, so far in 2014 we have observed a large number of web blogs written by SQL-on-Hadoop vendors that compare the performance of their system against the competitor systems reporting results using components of database benchmarks such as TPC-H [27] and TPC-DS [23]; Two benchmarks that are very popular for testing SQL-based query processing capabilities of relational databases.

A closer look at these performance comparisons reveals that the rules of the benchmarks are typically not followed. As we will discuss in more detail in the following section, it is common for vendors to pick a subset of the queries of the benchmark and perform a comparison using only those. Secondly, the queries are modified because more often they only support a limited subset of the SQL standard. Finally, it is not clear whether and how well the competitor open-source system was tuned.

According to the TPC standards, each database vendor installs, tunes and performs a full run of the database benchmark according to the benchmark specification, on his system only and then produces a report that describes all the details. This report and the performance results are audited by an accredited

---

[1] http://www.tpc.org/information/about/documentation/tpc_policies_v6.0.htm#_Toc367096059.
[2] http://en.wikipedia.org/wiki/David_DeWitt.

TPC Auditor and are submitted to the TPC for certification. When a potential customer wants to compare various systems in the same category from different vendors, she can review the high level performance metrics, including price performance, the detailed underlying implementation details, and the auditor's certification letter indicating compliance with all benchmark requirements.

In this paper, we want to point out the fact that in the world of SQL over Hadoop rigorous scientific benchmarking has been replaced by unscientific comparisons in the name of marketing and we would like to draw attention to this problem. We believe new benchmarks are needed that test not only the traditional structured query processing using SQL, but also emphasize the unique features of the Hadoop ecosystem and emerging big data applications. We also emphasize the need for benchmark specifications and industry commitment in order to bring standardization in the SQL-on-Hadoop space. Finally, we present our proposal towards these objectives.

2 Current Practices for Reporting Performance Results

In this section, we provide some examples of how SQL-on-Hadoop vendors use the TPC benchmarks when they evaluate their systems against the competitor systems.

One commonly used benchmark by all SQL-on-Hadoop vendors is the traditional TPC-DS benchmark [23]. The TPC-DS benchmark consists of 99 queries which access 7 fact tables and multiple dimension tables. TPC-H is also another popular benchmark used by the SQL-on-Hadoop vendors to test the performance of their systems. For those interested, the 22 TPC-H queries for Hive and Impala are available online [28, 29].

Recently, the Cloudera Impala developers have published a comparison between Hive, Impala and a traditional DBMS (called DBMS-Y) using TPC-DS as the workload basis [6, 24] and subsequent comparisons between Impala and other SQL-on-Hadoop systems [25, 26]. In their first comparison [24] they argue that using this workload Impala can be up to 69X faster than Hive and is generally faster than DBMS-Y with speedups up to 5X. In their subsequent comparison they find that Impala is on average 5X faster than the second fastest SQL-on-Hadoop alternative (Shark). The Impala developers have also provided the data definition language statements and queries that they have used in their study [13].

By taking a closer look at these queries, we will observe that only 19 queries out of the 99 TPC-DS queries are used. An additional query which is not part of the TPC-DS benchmark has also been introduced. Moreover, these queries access a single fact table only (out of the 7 fact tables that are part of the TPC-DS dataset). This results in query plans of a similar pattern when the fact table is joined with multiple dimension tables: the small dimension tables are broadcast to the nodes where the fact table resides and a join is performed locally on each of these nodes, without requiring any repartitioning/shuffling of the fact table's data.

Clearly, picking a subset of the queries that have a common processing pattern (for which a particular system is optimized) and testing the systems on only those queries do not reveal the full strengths and limitations of each system and is against the rules of the TPC-DS benchmark.

Another observation on the study of the queries in [24–26, 28, 29] is that some of these queries were modified in various ways:

- Cloudera's Impala does not currently support windowing functions and rollup. Thus, whenever a TPC-DS query included these features, these were removed from the query. This is not fair to other systems such as Apache Hive and possibly DBMS-Y which already support these features.
- An extra partitioning predicate on the fact table has beed added in the where clause of each query [24–26]. This predicate reduces the amount of the fact table data that need to be accessed during the query processing. It is worth noting that there exist advanced query rewrite techniques that introduce the correct partitioning predicates on the fact table without manual intervention [21].
- According to the TPC-DS specification the values in the query predicates change with the TPC-DS scale factor. However, the queries published in [13] do not use the correct predicate values for the supported scale factor.
- In both the TPC-H and the TPC-DS benchmarks, the queries contain predicates on the DATE attributes of the tables. These predicates typically select a date interval (e.g., within a month from a given date). This interval should be computed by the system under test by using, for example, built-in date functions. However, in the published TPC-H and TPC-DS inspired queries by the open-source vendors in [24–26, 28, 29], the date intervals are already pre-computed by the authors of the queries.
- In some queries, the predicates in the WHERE clauses are transformed into predicates in the ON clauses and are manually pushed down closer to the join operations.
- Some queries are re-written in a specific way that enforces the join ordering during the query execution.

When vendors present performance comparisons between different versions of their own systems, they use a very small set of the queries. This creates a false impression of a general characteristic when only a small biased (hand-picked) number of queries is used to substantiate a claim. For example, in a blog from Hortonworks [14], a single TPC-DS query is used to show the performance benefits using Hive's ORC file format with predicate pushdown over the previous Hive versions.

In this blog, Hortonworks also claims that the ORC columnar format results in better compression ratios than Impala's Parquet columnar format for the TPC-DS dataset at 500 scale factor. This claim, was later dismissed by Cloudera [6], which showed that if the same compression technique is used, the Parquet columnar format is more space-efficient than the ORC file format for the same dataset. This incident points out the importance of transparency. It is also worth noting that a 500 scale factor is not a valid scale factor for the TPC-DS workload.

It is quite clear that the rules established by the TPC for the TPC-H and the TPC-DS benchmark specifications are not followed in today's experimental comparisons for SQL-on-Hadoop systems. This can be quite misleading, since each SQL-on-Hadoop vendor can now modify, add or remove queries when comparing to other SQL-on-Hadoop systems. In fact, it is not even clear how well the competitor systems are tuned for these modified workloads.

Another performance comparison between different SQL-on-Hadoop systems have been published from the UC Berkeley AMPlab [1]. The benchmark used in these comparisons is inspired by the queries used in a 2009 SIGMOD paper [16] and contains only 4 queries. We would like to point out that the queries used in [16] were created in order to compare the vanilla MapReduce framework with parallel databases, and to prove that parallel databases excel at join. Since at that time MapReduce had limited functionality, these queries were very simple (e.g., contained at most one join operation). Today's SQL-on-Hadoop systems are much more sophisticated than the vanilla MapReduce framework and thus should not be evaluated with such simple benchmarks but with more sophisticated benchmarks that will reveal their full strengths and limitations.

We believe that these few examples clearly demonstrate the chaotic situation that currently exists in the SQL-on-Hadoop world when it comes to benchmarking and performance evaluation.

3   Whither SQL-on-Hadoop Benchmarking

Given the fact that benchmarking in the SQL-on-Hadoop world is in a "wild west" state, a natural question to ask is "What is the solution to this problem?" We believe that the SQL-on-Hadoop vendors should do the following:

1. Build on the decades of RDBMS benchmarking experience and move to a new generation of SQL over Hadoop benchmarking (e.g., [4]).
2. Employ good scientific experimental design and procedures that generate reproducible, comparable and trustworthy results.
3. Adhere to the TPC specifications and policies when using the TPC benchmarks.
4. Create new benchmarks for the SQL-on-Hadoop systems that represent the characteristics and features of the new generation of big data applications (e.g., BigBench [10], TPC-DS Hadoop/Hive Friendly, etc.)
5. Agree on the rules for new benchmarks, and extend existing ones as needed so that all vendors follow them when reporting benchmark results.

Since the benchmarking rules and policies are critical in bringing standardization in the SQL-on-Hadoop space, in the following sections, we discuss the benchmarking methodologies of different industry standard benchmark councils, as well as their relevance to the SQL-on-Hadoop benchmarking space. Finally, we present an outline of our own proposal for bringing standardization in this space.

3.1   TPC

The TPC [22] has been the leading benchmarking council for transaction processing and database benchmarks. However, TPC has been somewhat mired in the
traditions of the past, and has been slow to evolve and invent new benchmarks
to represent modern workloads and systems. While there has been movement to
providing more artifacts to lower the cost of participation, the benchmark kits
of the standard TPC benchmarks produced by each vendor remain proprietary.
The costs to audit and publish can be prohibitively high and are likely a key
reason for low vendor participation. This raises the issue on whether TPC is
the organization that will address the problems faced in the SQL-over-Hadoop
space.

Recently, the TPC has started re-inventing itself by introducing the TPC
Express process [9] and by launching TPCx-HS [30], the first benchmark that
follows the TPC Express process. The TPC Express process aims at lowering
the cost of participation at TPC and at making the TPC benchmarks accessible
to a broad class of practitioners including academia, consumers, analysts and
computer hardware and software manufacturers [9]. The TPCx-HS benchmark
is the first TPC benchmark focused on big data systems such as Hadoop. As
opposed to the previous TPC benhmark, the TPCx-HS benchmark is available
via the TPC Web site in the form of a downloadable kit. The existence of a kit,
independent of the vendor that runs the benchmark, is a key characteristic of the
TPC Express process which aims to make the benchmark more readily available.
As opposed to other database benchmarks such as TPC-H, whose results must
be validated by a certified TPC auditor, the TPCx-HS benchmark results can
also be validated using a peer-review process. The members of the peer-review
committee are official TPCx-HS members.

Another problem of the TPC benchmarks is that they have historically taken
a rather narrow view of the "measure of goodness" while real customers have
broader considerations. The former CTO of SAP, Vishal Sikka, published a blog
posting that attempted to articulate the characteristics that should be considered
in a new benchmark[3]. While the blog itself was a thinly disguised advertisement
for HANA, Vishal does raise some interesting aspects that can and should be
considered when constructing a new benchmark. In particular, he points to five
core dimensions that should be considered:

1. going deep (the benefit of allowing unrestricted query complexity)
2. going broad (the benefit of allowing unrestricted data volume and variety)
3. in real-time (the benefit of including the most recent data into the analysis)
4. within a given window of opportunity (the benefit of rapid response time)
5. without pre-processing of data (the cost of data preparation)

Most of these are relevant to current and future Hadoop systems and should
be taken into account when defining a new benchmark for the SQL-on-Hadoop
systems.

---

[3] http://www.saphana.com/community/blogs/blog/2013/09/16/
does-the-world-need-a-new-benchmark.

3.2 SPEC

The Standard Performance Evaluation Corporation (SPEC) [19] is formed to establish, maintain and endorse a standardized set of relevant benchmarks for high-performance computers. SPEC also reviews and publishes submitted results from the SPEC member organizations and other benchmark licensees. SPEC employs a peer-review scheme of the benchmarking results (instead of an auditing scheme) that has been quite successful. The members of the peer-review committee are official SPEC members. Through this membership, the companies that are willing to accept SPEC's standards can also participate in the development of the benchmarks.

3.3 STAC

The Securities Technology Analysis Center (STAC) [20] is a relatively new organization. It first formed a Benchmark Council in 2007. It is a large, well-funded organization consisting of over 200 financial institutions and 50 vendor organizations, focused on the securities industry. One unique aspect of this organization is that it is run by companies representing "consumers" of IT products. Vendors, while encouraged to participate, do not control the organization. STAC has only recently become interested in Big Data and formed a Big Data special interest group in 2013. The STAC members have written a white paper that characterizes the major Big Data use cases they envision in the securities and banking industry [7].

STAC has begun working on a Big Data benchmark specification whose details are restricted to its members. It is too early to know how strong a benchmark will emerge, and what value it will have outside the securities industry. We encourage the Big Data community to stay abreast of developments at STAC, and encourage STAC to be more open about the benchmarks and specifications that it generates.

4 Our Proposal

The rate of change, the number of new players, and the industry-wide shift to new communication modes (e.g. blogs, tweets) make it next to impossible to conduct benchmarks using the traditional auditing procedures (such as the TPC auditing process) using database TPC benchmarks. Our belief is that the SQL-on-Hadoop community should build on the experiences of different industry standard benchmark councils to bring standardization in the SQL-on-Hadoop space and to produce fair and meaningful benchmarking results. The necessary steps needed towards this goal are the following:

1. Robust benchmark specifications
2. Flexible, portable, and easy to use benchmarking kits
3. Cost-effective, timely, efficient yet high quality peer-reviewing procedures

We believe that the existence of flexible and downloadable benchmarking kits is a significant step towards the widespread adoption of any benchmark. Proprietary benchmark kits result in high implementation cost, and thus make the benchmarking process expensive to start in the first place. This observation has already been made by TPC, and thus the TPCx-HS benchmark is the first TPC benchmark that is vendor-neutral: It comes with a downloadbale benchmarking kit. This is a key characteristic of the TPC Express process.

Regarding the reviewing of the benchmarking results, we need a robust scheme that will bring organization, transparency and objectivity in the SQL-on-Hadoop benchmarking space. We believe that the best solution to the SQL-on-Hadoop world's chaotic state is the use of a peer-review approach. The goal of this approach is to develop timely, cost-effective but high quality "reviews" to minimize the "marketing" effects. Using this approach, every vendor that publishes or compares a set of SQL-on-Hadoop systems, writes a performance report that includes a description of the hardware and software configuration and the tuning process. This report is peer-reviewed, not only by the vendors of the systems tested but by other independent reviewers in the industrial or academic setting. Once an agreement is reached the results would be published online. As noted in an earlier section, the peer-review scheme has already been used by organizations such as SPEC [19] and has been quite succesful.

We believe that all the SQL-on-Hadoop vendors would agree on this approach, since they will be able to argue about the performance of their systems whenever another vendor conducts a performance comparison that incorporates their system. The most challenging part of this process is to ensure that the SQL-on-Hadoop vendors be prevented from blocking the publication of unfavorable results for their systems. To avoid such cases, we propose a "revision" process in which the SQL-on-Hadoop vendors that doubt the validity of the results, will provide concrete feedback that proposes configuration changes that should to be made. A maximum number of revision requests per vendor could also be set (e.g., up to two revision requests). This approach guarantees that: (a) the vendors that have doubts about the validity of the experimental setting/result will be forced to provide accurate and detailed feedback and (b) these vendors will be prevented from indefinitely blocking the publication of the results.

5   Conclusions

The existence of new, more sophisticated benchmarks that can represent the new generation workloads, is certainly a big step for the SQL-on-Hadoop community. However, their formulation is not going to bring any standardization unless these benchmarks are accompanied by rules and policies that will ensure transparency and objectivity. Otherwise, these benchmarks will be abused in the name of marketing, similar to what is happening now with the existing TPC benchmarks. To realize this vision, all the SQL-on-Hadoop vendors should come to an agreement on how to use the benchmarks, and how to report performance results using them. We believe that a peer-review approach along with the existence of portable and easy-to-use benchmarking kits is the only viable solution. Furthermore, templates for summarizing results in a standard way, similar to a TPC Executive Summary, should be created and provided to all. Of course, we do not expect that the existence of published unofficial results that present performance evaluations for different systems will cease to exist. However, if the community agrees upon the standards, publishes results based on the standards, and uses an effective reviewing scheme, then the importance of these web blogs and their impact on the end-user will be significantly reduced. This is exactly the same reason that led to the formation of the TPC and its first benchmark, TPC-A, more than 25 years ago. IBM is eager to join the community in bringing order to the exciting world of SQL-on-Hadoop benchmarking.

References

1. AMPLAB Big Data Benchmark. https://amplab.cs.berkeley.edu/benchmark/
2. Apache Hive. http://hive.apache.org/
3. IBM InfoSphere BigInsights. http://www-01.ibm.com/support/knowledgecenter/ SSPT3X_3.0.0/com.ibm.swg.im.infosphere.biginsights.product.doc/doc/whats new.html
4. Chen, Y., Raab, F., Katz, R.: From TPC-C to big data benchmarks: a functional workload model. In: Rabl, T., Poess, M., Baru, C., Jacobsen, H.-A. (eds.) WBDB 2012. LNCS, vol. 8163, pp. 28–43. Springer, Heidelberg (2014)
5. Cloudera Impala. http://www.cloudera.com/content/cloudera/en/products-and-services/cdh/impala.html
6. Cloudera Impala Technical Deep Dive. http://www.slideshare.net/huguk/hug-london2013
7. Costley, J., Lankford, P.: Big Data Cases in Banking and Securities (2014). https:// stacresearch.com/news/2014/05/30/big-data-use-cases-banking-and-securities
8. DeWitt, D.J., Nehme, R.V., Shankar, S., Aguilar-Saborit, J., Avanes, A., Flasza, M., Gramling, J.: Split query processing in polybase. In: ACM SIGMOD, pp. 1255–1266 (2013)
9. TPC Express. http://www.tpc.org/tpctc/tpctc2013/slides_and_papers/004.pdf
10. Ghazal, A., Rabl, T., Hu, M., Raab, F., Poess, M., Crolotte, A., Jacobsen, H.-A.: BigBench: towards an industry standard benchmark for big data analytics. In: ACM SIGMOD, pp. 1197–1208 (2013)
11. Gray, J. (ed.): The Benchmark Handbook for Database and Transaction Systems, 2nd edn. Morgan Kaufmann, San Francisco (1993). http://research.microsoft.com/ en-us/um/people/gray/benchmarkhandbook/chapter2.pdf
12. Groves, T.: The Big Deal about InfoSphere BigInsights v3.0 is Big SQL. http:// www.ibmbigdatahub.com/blog/big-deal-about-infosphere-biginsights-v30-big-sql
13. Impala TPC-DS Kit. https://github.com/cloudera/impala-tpcds-kit
14. ORCFile in HDP 2.0. http://hortonworks.com/blog/orcfile-in-hdp-2-better-compression-better-performance/
15. Ozcan, F., Harris, S.: Blistering Fast SQL Access to Your Hadoop Data. http:// www.ibmbigdatahub.com/blog/blistering-fast-sql-access-your-hadoop-datal
16. Pavlo, A., Paulson, E., Rasin, A., Abadi, D.J., DeWitt, D.J., Madden, S., Stonebraker, M.: A comparison of approaches to large-scale data analysis. In: ACM SIGMOD. ACM, New York (2009)

17. Pivotal HAWQ. http://pivotalhd.docs.gopivotal.com/getting-started/hawq.html
18. Presto. http://prestodb.io/
19. SPEC: Standard Performance Evaluation Corporation. http://www.spec.org/
20. STAC: Security Technology Analysis Center. https://stacresearch.com/
21. Szlichta, J., Godfrey, P., Gryz, J., Ma, W., Pawluk, P., Zuzarte, C.: Queries on dates: fast yet not blind. In: Proceedings of the 14th International Conference on Extending Database Technology, EDBT/ICDT 2011, pp. 497–502. ACM, New York (2011)
22. Transaction Processing Performance Council. http://www.tpc.org
23. The TPC-DS Benchmark. http://www.tpc.org/tpcds/
24. TPC-DS-like Workload on Impala (part 1). http://blog.cloudera.com/blog/2014/01/impala-performance-dbms-class-speed/
25. TPC-DS-like Workload on Impala (part 2). http://blog.cloudera.com/blog/2014/05/new-sql-choices-in-the-apache-hadoop-ecosystem-why-impala-continues-to-lead/
26. TPC-DS-like Workload on Impala (part 3). http://blog.cloudera.com/blog/2014/09/new-benchmarks-for-sql-on-hadoop-impala-1-4-widens-the-performance-gap/
27. The TPC-H Benchmark. http://www.tpc.org/tpch/
28. TPC-H Scripts for Hive. https://issues.apache.org/jira/browse/HIVE-600
29. TPC-H Scripts for Impala. https://github.com/kj-ki/tpc-h-impala
30. The TPCx-HS Benchmark. http://www.tpc.org/tpcx-hs/spec/tpcx-hs-specification-v1.1.pdf

The Emergence of Modified Hadoop Online-Based MapReduce Technology in Cloud Environments

Shaikh Muhammad Allayear[1(✉)], Md. Salahuddin[1], Delwar Hossain[1], and Sung Soon Park[2]

[1] Department of Computer Science and Engineering, East West University, Dhaka, Bangladesh
allayear@ewubd.edu, 2010-2-60-001@ewu.edu.bd, delwarhossain518@live.com
[2] Anyang University, Anyang, South Korea
sspark@anyang.ac.kr

Abstract. The exponential growth of data first presented challenges to cutting-edge businesses such as Goggle, Yahoo, Amazon, Microsoft, Facebook, and Twitter. Data volumes to be processed by cloud applications are growing much faster than computing power. This growth demands new strategies for processing and analyzing information. Hadoop MapReduce has become a powerful computation model that addresses those problems. MapReduce is a programming model that enables easy development of scalable parallel applications to process vast amounts of data on large clusters. Through a simple interface with two functions, map and reduce, this model facilitates parallel implementation of many real world tasks such as data processing for search engines and machine learning. Earlier versions of Hadoop MapReduce had several performance problems like connection between map to reduce task, data overload and slow processing. In this paper, we propose a modified MapReduce architecture – MapReduce Agent (MRA) – that resolves those performance problems. MRA can reduce completion time, improve system utilization, and give better performance. MRA employs multi-connection which resolves error recovery with a Q-chained load balancing system. In this paper, we also discuss various applications and implementations of the MapReduce programming model in cloud environments.

1 Introduction

Cloud computing is the delivery of computing services over the Internet. Cloud services allow individuals and businesses to use software and hardware that are managed by third parties at remote locations. Examples of cloud services include online file

This research (Grants NO. 2013-140-10047118) was supported by the 2013 Industrial Technology Innovation Project Funded by Ministry Of Science, ICT and Future Planning.
The source code for HOP can be downloaded from http://code.google.com/p/hop.

© Springer International Publishing Switzerland 2015
T. Rabl et al. (Eds.): WBDB 2014, LNCS 8991, pp. 73–86, 2015.
DOI: 10.1007/978-3-319-20233-4_8

storage, social networking sites, webmail, and online business applications. The cloud computing model allows access information and computer resources from anywhere where a network connection is available. To make full use of big data, tens of terabytes (TBs) or tens of petabytes (PBs) of data need to be handled. To process vast amounts of data Hadoop MapReduce is the basic technologies for big data processing in cloud environments. Google proposed MapReduce. The MapReduce framework simplifies the development of large-scale distributed applications on clusters of commodity machines. MapReduce is typically applied to large batch-oriented computations that are concerned primarily with time to job completion. The Google MapReduce framework [1] and open-source Hadoop system reinforce this usage model through a batch-processing implementation strategy: the entire output of each map and reduce task is materialized to a local file before it can be consumed by the next stage. Materialization allows for a simple and elegant checkpoint/restart fault tolerance mechanism that is critical in large deployments, which have a high probability of slowdowns or failures at worker nodes and traditional MapReduce have some limitation like performance problem, connection problem etc.

To solve the above discussed problems we propose a modified MapReduce architecture that is MapReduce Agent (MRA). MRA provides several important advantages to the MapReduce framework. We highlight the potential benefits first:

- In the MapReduce framework, data is transmitted from the map to the reduce stage. So there may be connection problem. To solve this problem MRA creates iSCSI [2] Multi-Connection and Error Recovery Method [3] to avoid drastic reduction of transmission rate from TCP congestion control mechanism and guarantee fast retransmission of corruptive packet without TCP re-establishment.
- For fault tolerance and workload, MRA creates Q-chained cluster. A Q-chained cluster [3] is able to balance the workload fully among data connections in the event of packet losses due to bad channel characteristics.
- In Cloud computing environment a popular data processing engine for big data is Hadoop MapReduce due to ease-of-use, scalability, and failover properties.

The rest of this paper is organized as follows. Overview of the big data, cloud computing, iSCSI protocol, Hadoop MapReduce architecture and pipelining mechanism [5] are described in Sect. 2. In Sect. 3, we describe our research motivations. We describe our proposed model of MapReduce Agent briefly in Sect. 4. We evaluate the performance and discuss results in Sect. 5. Finally in Sect. 6, we provide the conclusion of this paper.

2 Background

In this section, besides the iSCSI protocol we review the big data implementation in the cloud environment, MapReduce programming model and describe the salient features of Hadoop, a popular open-source implementation of MapReduce.

2.1   Cloud Computing

Cloud computing is an emerging technology for large scale data analysis, providing scalability to thousands of computers, in addition to fault tolerance and cost effectiveness. Generally, a cloud computing environment is a large-scale distributed network system implemented based on a number of servers in data centers. The cloud services are generally classified based on a layer concept. In the upper layers of this paradigm, Infrastructure as a Service (IaaS), Platform as a Service (PaaS), and Software as a Service (SaaS) are stacked.

Infrastructure as a Service (IaaS): IaaS is built on top of the data center layer. IaaS enables the provision of storage, hardware, servers and networking components. The client typically pays on a per-use basis. The examples of IaaS are Amazon EC2 (Elastic Cloud Computing) and S3 (Simple Storage Service).

Platform as a Service (PaaS): PaaS offers an advanced integrated environment for building, testing and deploying custom applications. The examples of PaaS are Google App Engine, Microsoft Azure, and Amazon MapReduce/Simple Storage Service.

Software as a Service (SaaS): SaaS supports a software distribution with specific requirements. In this layer, the users can access an application and information remotely via the Internet and pay only for that they use. Salesforce is one of the pioneers in providing this service model. Microsoft's Live Mesh also allows sharing files and folders across multiple devices simultaneously.

2.2   Big Data Management System

Big data includes structured data, semi-structured (XML or HTML tagged text) and unstructured (PDF's, e-mails, and documents) data. Structured data are those data formatted for use in a database management system. Semi-structured and unstructured data include all types of unformatted data including multimedia and social media content. Hadoop, used to process unstructured and semi-structured big data, uses the MapReduce paradigm to locate all relevant data then select only the data directly answering the query. NoSQL, MongoDB, and TerraStore process structured big data.

2.3   Hadoop Architecture

Hadoop [4] includes Hadoop MapReduce, an implementation of MapReduce designed for large clusters, and the Hadoop Distributed File System (HDFS), a file system optimized for batch-oriented workloads such as MapReduce. In most Hadoop jobs, HDFS is used to store both the input to the map step and the output of the reduce step. Note that HDFS is not used to store intermediate results (e.g., the output of the map step): these are kept on each node's local file system.

A Hadoop installation consists of a single master node and many worker nodes. The master, called the Job-Tracker, is responsible for accepting jobs from clients, dividing those jobs into tasks, and assigning those tasks to be executed by worker nodes.

2.4   Map Task Execution

Each map task is assigned a portion of the input file called a split. By default, a split contains a single HDFS block (64 MB by default) [4], so the total number of file blocks determines the number of map tasks. The execution of a map task is divided into two phases (Fig. 1).

```
public interface Mapper < K1, V1, K2, V2>{
    void map(K1 key, V1 value, OutputCollector<K2, V2> output);
    void close ();
}
```

Fig. 1. Map function interface

- The map phase reads the task's split from HDFS, parses it into records (key/value pairs), and applies the map function to each record.
- After the map function has been applied to each input record, the commit phase registers the final output with the TaskTracker, which then informs the JobTracker that the task has finished executing.

2.5   Reduce Task Execution

The execution of a reduce task is divided into three phases (Fig. 2).

```
public interface Reducer<K2, V2, K3, V3>{
    void reducer (K2 key,Iterator<V2> values, OutputCollector<K3, V3> output);
    void close();
}
```

Fig. 2. Reduce function interface.

- The shuffle phase fetches the reduce task's input data. Each reduce task is assigned a partition of the key range produced by the map step, so the reduce task must fetch the content of this partition from every map task's output.
- The sort phase groups records with the same key together.
- The reduce phase applies the user-defined reduce function to each key and corresponding list of values.

2.6   Pipelining Mechanism

In pipelining version of Hadoop [5], they developed the Hadoop online prototype (HOP) that can be used to support continuous queries: MapReduce jobs that run continuously. They also proposed a technique known as online aggregation which can provide initial estimates of results several orders of magnitude faster than the final results. Finally the pipelining can reduce job completion time by up to 25 % in some scenarios.

2.7 iSCSI Protocol

iSCSI (Internet Small Computer System Interface) is a transport protocol that works on top of TCP [2]. iSCSI transports SCSI packets over TCP/IP. iSCSI client-server model describes clients as iSCSI initiator and data transfer direction is defined with regard to the initiator. Outbound or outgoing transfers are transfer from initiator to the target.

3 Motivations

When a company needs to store and access more data it has multiple choices. One option would be to buy a bigger machine with more CPU, RAM, disk space, etc. This is known as scaling vertically. Of course, there is a limit to the size of single machine that are available and at internet scale this approach is not viable. Another option is cloud computing, here we can store more data. In Cloud computing environment a popular data processing engine for big data is Hadoop MapReduce.

In Cloud computing environment various cloud clients store and process data in a wireless network that has important matter to think about total performance during data sending and receiving. As we know that in wireless network there bandwidth is narrow so during packet exchange time there is huge data overheads. So data sending and receiving parameters needs tuning to be optimized unnecessary packet exchange. Our proposed method is offering to remove unnecessary packet exchange of an iSCSI protocol and to reduce a network overhead.

In the pipelining mechanism of Hadoop MapReduce a naïve implementation is used to send data directly from map to reduce tasks using TCP [5]. When a client submits a new job to Hadoop, the JobTracker assigns the map and reduce tasks associated with the job to the available TaskTracker slots. In the modified Hadoop system each reduce task contacts every map task upon initiation of the job and opens a TCP socket, which will be used to send the output of the map function. The drawback of this solution is that TCP congestion during data transmission cab occur. In this case, TCP connections are being disconnected and after that data must be retransmitted, which takes a long time. To solve this problem, we propose MRA that can send data without retransmission using iSCSI multi-connection and also manages load balancing of data because iSCSI protocol works over TCP. Another motivation is that the iSCSI protocol is based on block I/O and Hadoop's map task also assigns HDFS blocks for the input process.

4 Proposed Model: MapReduce Agent

In cloud computing environments a popular data processing engine for big data is Hadoop-MapReduce due to its ease-of-use, scalability, and failover properties. But traditional MapReduce sometimes has poor performance due to connection problems and slow processing. To resolve those problems and improve the limitations of Hadoop MapReduce, we create MRA, which can improve the performance. Traditional MapReduce implementations also provides a poor interface for interactive data analysis, because they do not emit any output until the map task has been executed to completion. After producing the output of the map function, MRA creates multi-connections with the reducer rapidly (Fig. 3).

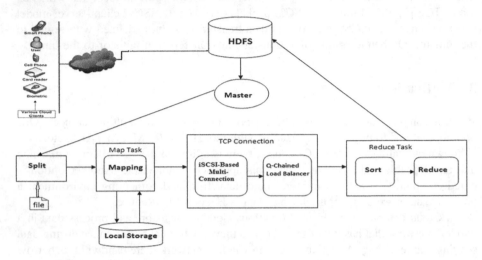

Fig. 3. Proposed MapReduce mechanism to process data in cloud environment.

If one connection falls or a data overload problem occurs then the rest of job will distributed to other connections. Our Q-Chained cluster load balancer does this job [3]. So that the reducer can continue its work, which reduces the job completion time (Fig. 4).

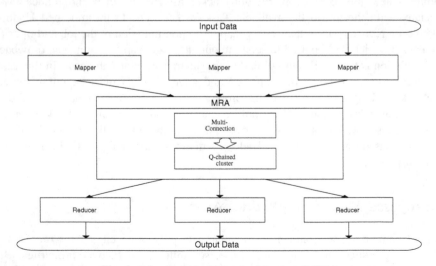

Fig. 4. Map Reduce Agent Architecture (MRA).

4.1 Multi-Connection and Error Recovery Method of ISCSI

In order to alleviate the degradation of the iSCSI-based remote transfer service caused by TCP congestion control, we propose the MRA Multi-Connection and Error Recovery method for one session, which uses multiple connections for each session. As mentioned in [3], in a single TCP network connection when congestion occurs by a timeout or the reception of duplicate ACKs (Acknowledgement) then one half of the current window size is saved in sstresh (slow start window). Additionally, if the congestion is indicated by a timeout, cwnd (congestion window) is set to one segment. This may cause a significant degradation in online MapReduce performance. On the other hand in the Multi-Connection case, if TCP congestion occurs within a connection, the takeover mechanism selects another TCP connection.

The general overview of the proposed Multi-Connection and Error Recovery based on iSCSI protocol scheme, which has been designed for iSCSI based transfer system. When the mapper (worker) is in active mode or connected mode for reduce job that time session is started. This session is indicated to be a collection of multiple TCP connection. If packet losses occur due to bad channel characteristics in any connection, our proposed scheme will pick out Q-Chained Cluster's balanced redistribute data by the other active connections.

We use a filter mechanism to control parameter. Figure 5 shows the parameter filter module, which checks the network status, and calculates the channel number, which is best suited for network resource. The filter also filters the parameter of iSCSI initiator and target. In iSCSI remote storage systems there are also device commands and iSCSI commands. The filter module checks the commands and network status of both initiator and target. If the command parameter carries the device command then it sends to the iSCSI device directly and if the command parameter is iSCSI command related like NOP IN, NOP OUT [7] then it does not need to send them to the device controller.

This way we can reduce the network overhead and increase the iSCSI remote storage system performance. The parameter controller measures the Round-Trip Time (RTT) in TCP two-way handshake to determine the appropriate number of TCP connections for a specific destination.

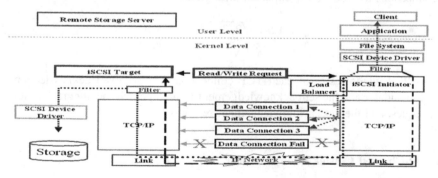

Fig. 5. Overview of Multi-connection and Error Recovery Method of iSCSI [3].

4.1.1   TCP Two Way Handshaking

With RTT we can measure TCP two-way handshaking between the initiator and the target, it can be more efficient at avoiding filtering and inflation of packets than ICMP probes. The multi-TCP connections controller negotiates the number of connections between the initiator and the target for data transmission, according to Eq. (2) using parameter (RTT), which was collected by the parameter collector.

As given: p is a packet drop rate. T bps: the maximum sending rate for a TCP connection. B (bytes): TCP connection sending packets with a fairly constant RTT of R seconds. Given the packet drop rate p, the minimum Round-trip time R, and the maximum packet size B, we can use the Eq. (1) to calculate the maximum arrival rate from a conformant TCP connection.

$$T \leq \frac{1.5 * \sqrt{2/3 * B}}{R * \sqrt{p}} \tag{1}$$

Equation (2) shows that the number of established TCP connections (N) used in Multi-Connection iSCSI depends on RTT (Rt) measured by the parameter collector. The minimum RTT can determine the number of connections to be opened as shown in Fig. 6.

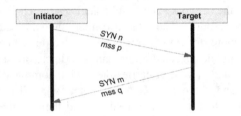

Fig. 6. Two-way handshaking of TCP Connection.

However, while the use of concurrent connections increases throughput, it also increases the packet drop rate. Therefore, it is important to obtain the optimal number of connections to produce the expected throughput.

$$T \leq \frac{1.5 * \sqrt{2/3 * B}}{R * \sqrt{p}} \leq \frac{N * W}{Rt} \tag{2}$$

Where W is window size of each TCP connection. The received acknowledgement from the initiator and the sent acknowledgement from target will be integrated with the next request and the next acknowledgement response.

4.2   Q-Chained Cluster Load Balancer

Q-chained cluster is able to balance the workload fully among data connections in the event of packet losses due to bad channel characteristics. When a congestion occurs in a data connection, this module can do a better job of balancing the workload, which is originated by congestion connection. It will be distributed among N-1 connections instead of a single data connection. Figure 7 illustrates how the workload is balanced in the event of congestion occurrence in a data connection (Data Connection 1 in this example) with Q-chained cluster. For example, with the congestion occurre nce of Data Connection 1, the primary data Q1 is no longer transmitted in the congested connection for the TCP input rate to be throttled and thus its recovery data Q1 of Data Connection 1 is passed to Data Connection 2 for conveying storage data. However, instead of requiring Data Connection 2 to process all data both Q2 and Q1, Q-chained cluster offloads 4/5ths of the transmission of Q2 by redirecting them to Q2 in Data Connection 3. In turn, 3/5ths of the transmission of Q3 in Data Connection 3 are sent to Q3. This dynamic reassignment of the workload results in an increase of 1/5th in the workload of each remaining data connection.

Data connection	0	1	2	3	4	5
Primary Data	Q_0	F	$1/5Q_2$	$2/5Q_3$	$3/5Q_4$	$4/5Q_5$
Recovery Data	$1/5q_5$	F	q_1	$4/5q_2$	$3/5q_3$	$2/5q_2$

Fig. 7. Q-Chained load balancer.

5 Performance Evaluation

As per as [5] we also evaluate the effectiveness of online aggregation, we performed two experiments on Amazon EC2 using different data sets and query workloads. In their first experiment [5], the authors wrote a "Top-K" query using two MapReduce jobs: the first job counts the frequency of each word and the second job selects the K most frequent words. We ran this workload on 5.5 GB of Wikipedia article text stored in HDFS, using a 128 MB block size. We used a 60-node EC2 cluster; each node was a "high-CPU medium" EC2 instance with 1.7 GB of RAM and 2 virtual cores. A virtual core is the equivalent of a 2007-era 2.5 Ghz Intel Xeon processor. A single EC2 node executed the Hadoop Job- Tracker and the HDFS NameNode, while the remaining nodes served as slaves for running the TaskTrackers and HDFS DataNodes.

A thorough performance comparison between pipelining, blocking and MRA is beyond the scope of this paper. In this section, we instead demonstrate that MRA can reduce job completion times in some configurations. We report performance using both large (512 MB) and small (32 MB) HDFS block sizes using a single workload (a wordcount job over randomly-generated text). Since the words were generated using a uniform distribution, map-side combiners were ineffective for this workload. We performed all experiments using relatively small clusters of Amazon EC2 nodes. We also did not consider performance in an environment where multiple concurrent jobs are executing simultaneously.

5.1   Performance Results of ISCSI Protocol for Multi-Connection

Experimental Methodology:

Our scheme's throughput in different RTTs are measured for different numbers of connections in Fig. 8. We see the slowness of the rising rate of throughput between 8 connections and 9 connections. This shows that reconstructing the data in turn influences throughputs and the packet drop rates are increased when the number of TCP connections is 9 as the maximum use of concurrent connections between initiator and target.

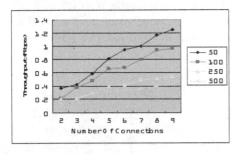

Fig. 8. Throughput of Multi-Connection iSCSI System. Y axis is containing throughput easurement with Mbps & X axis is for number of connections. 50, 100, 250 and 500 RTT are measured by ms.

Fig. 9. Throughput of Multi-Connection iSCSI vs iSCSI at different error rates. Y axis is throughput & X axis is for bit error rate.

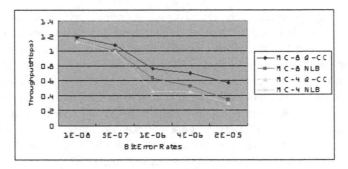

Fig. 10. Q-Chained cluster load balancer vs No load balancer. MC: Multi Connection, Q-CC: Q-Chained cluster NLB: No load balancer.

Therefore, 8 is the maximum optimal number of connections from a performance point of view. Multi-Connection iSCSI mechanism also works effectively because the data transfer throughputs increase linearly when the round trip time is larger than 250 ms.

In Fig. 9, the performance comparison of Multi-Connection iSCSI and iSCSI at different bit-error rates is shown. We see that for bit-error rates of over 5.0×10-7the Multi-Connection iSCSI (2 connections) performs significantly better than the iSCSI (1 connection), achieving a throughput improvement about 24 % in SCSI read.

Moreover, as bit-error rates go up, the figure shows that the rising rate of throughput is getting higher at 33 % in 1.0 × 10-6, 39.3 % in 3.9 × 10-6and 44 % in 1.5 × 10-5. Actually, Multi-Connection iSCSI can avoid the forceful reduction of transmission rate efficiently from TCP congestion control using another TCP connection opened during a service session, while iSCSI does not make any progress. Under statuses of low bit error rates (< 5.0 × 10-7), we see little difference between Multi-Connection iSCSI and iSCSI. At such low bit errors iSCSI is quite robust at handling these.

In Fig. 10, Multi-Connection iSCSI (8 connections) with Q-Chained cluster shows average performance improvement of about 11.5 %. It can distribute the workload among all remaining connections when packet losses occur in any connection. To recall an example given earlier, with M = 6, when congestion occurs in a specific connection, the workload of each connection increases by only 1/5. However, if Multi-Connection iSCSI (proposed scheme) establishes a performance baseline without load balancing, any connection, which is randomly selected from takeover mechanism, is overwhelmed.

5.2 Experimental Results of TCP Two Way Handshaking

Comparing two way handshaking method with three way handshaking, we have achieved better performance which is shown in Fig. 11.

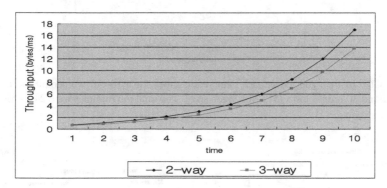

Fig. 11. Comparison of data transmission, in between two-way handshaking and three-way handshaking mode.

5.3 Performance Results and Comparison on MapReduce

In the Hadoop map reduce architecture [4, 5]; their first task is to generate output which is done by map task consume the output by reduce task. The whole thing makes the process lengthy because reduce tasks have to wait for the output of the map tasks. Using pipelining mechanism [5], they send output of map task immediately after generation of per output to the reduce task so it takes less time than Hadoop Ma-pReduce. During the transmission (TCP) if any problem occurred then they retransmit again which takes more time and drastically reduces the performance of the MapReduce mechanism (Figs. 12, 13, 14, 15, 16, 17).

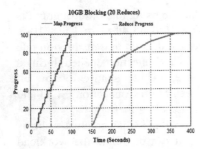

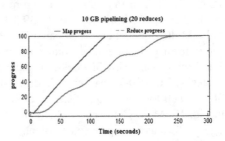

Fig. 12. CDF of map and reduce task completion times for a 10 GB wordcount job using 20 map tasks and 20 reduce tasks (512 MB block size). The total job runtimes were 361 s for blocking.

Fig. 13. CDF of map and reduce task completion times for a 10 GB wordcount job using 20 map tasks and 20 reduce tasks (512 MB block size). The total job runtimes were 290 s for pipelining.

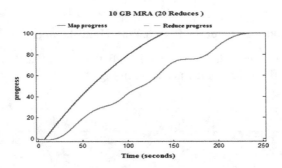

Fig. 14. CDF of map and reduce task completion times for a 10 GB wordcount job using 20 map tasks and 20 reduce tasks (512 MB block size). The total job runtimes were 240 s for MRA.

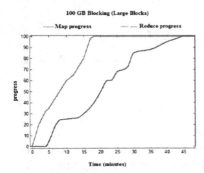

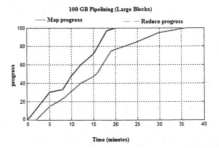

Fig. 15. CDF of map and reduce task completion times for a 100 GB wordcount job using 240 map tasks and 60 reduce tasks (512 MB block size). The total job runtimes were 48 min for blocking.

Fig. 16. CDF of map and reduce task completion times for a 100 GB wordcount job using 240 map tasks and 60 reduce tasks (512 MB block size). The total job runtimes were 36 min for pipelining.

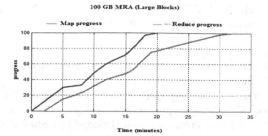

Fig. 17. CDF of map and reduce task completion times for a 100 GB wordcount job using 240 map tasks and 60 reduce tasks (512 MB block size). The total job runtimes were 32 min for MRA.

On the other hand our proposed mechanism (MRA) recovers the drawback by using multi-connection and Q-chained load balancer method. In these circumstances MRA may prove its better time of completion.

6 Conclusion

Cloud technology progress & increased use of the Internet are creating very large new datasets with increasing value to businesses and processing power to analyze them affordable. The Hadoop-MapReduce programming paradigm has a substantial base in the big data community due to the cost-effectiveness on commodity Linux clusters and in the cloud via data upload to cloud vendors, who have implemented Hadoop/HBase.

Finally we can say that our proposed model MRA resolves all the limitation of Hadoop Map Reduce and it can reduce the time to job completion. Our modified MapReduce architecture that can play an important role to process big data in cloud environment efficiently.

References

1. Dean, J., Ghemawat, S.: MapReduce: Simplified dataprocessing on large clusters. In: OSDI (2004)
2. SAM-3 Information Technology – SCSI Architecture Model 3, Working Draft, T10 Project 1561-D, Revision7 (2003)
3. Allayear, S.M., Park, S.S.: iSCSI multi-connection and error recovery method for remote storage system in mobile appliance. In: Gavrilova, M.L., Gervasi, O., Kumar, V., Tan, C., Taniar, D., Laganá, A., Mun, Y., Choo, H. (eds.) ICCSA 2006. LNCS, vol. 3981, pp. 641–650. Springer, Heidelberg (2006)
4. Hadoop. http://hadoop.apache.org/mapreduce/
5. Condie, T., Conway, N., Alvaro, P., Hellerstein, J.M.: UC Berkeley: MapReduce Online. Khaled Elmeleegy, Russell Sears (Yahoo! Research)

6. Allayear, S.M., Park, S.S.: iSCSI protocol adaptation with NAS system via wireless environment. In: International Conference on Consumer Electronics (ICCE), Las Vegus, USA (2008)
7. RFC 3270. http://www.ietf.org/rfc/rfc3720.txt
8. Daneshyar, S., Razmjoo, M.: Large-Scale Data Processing Using Mapreduce in Cloud Computing Environment
9. Changqing Ji*†, Yu Li‡, Wenming Qiu‡, Uchechukwu Awada‡, Keqiu Li‡ : Big Data Processing in Cloud Computing Environments
10. Rabi Prasad Padhy: Big Data Processing with Hadoop-MapReduce in Cloud Systems
11. Chan, J.O.: An Architecture for Big Data Analytics
12. Hellerstein, J.M., Haas, P.J., Wang, H.J.: Online aggregation. In: SIGMOD (1997)
13. Caceres, R., Iftode, L.: Improving the performance of reliable transport protocols inMobile computing environments. IEEE JSAC
14. Laurila, J.K., Blom, J., Dousse, O., Gatica-Perez, D.: The Mobile Data Challenge: Big Data for Mobile Computing Research
15. Satyanarayanan, M.: Mobile computing: the next decade. In: Proceedings of the 1st ACM Workshop on Mobile Cloud Computing & Services: Social Networks and Beyond (MCS) June 2010
16. Verma, A., Zea, N., Cho, B., Gupta, I., Campbell, R.H.: Breaking the MapReduce Stage Barrier*
17. Stokely, M.: Histogram tools for distributions of large data sets
18. Lu, L., Shi, X., Jin, H., Wang, Q., Yuan, D., Wu, S.: Morpho: A decoupled MapReduce framework for elastic cloud computing
19. Hao, C., Ying, Q.: Research of Cloud Computing based on the Hadoop platform. Chengdu, China, pp. 181–184, 21-23 October 2011
20. Armbrust, M., Fox, A., Griffith, R., Joseph, A.D., Katz, R.H., Konwinski, A., Lee, G., Patterson, D.A., Rabkin, A., Stoica, I., Zaharia, M.: Above the Clouds: a Berkeley View of Cloud Computing, Tech. Rep., University of California at Berkeley (2009)
21. Palanisamy, B., Singh, A., Liu, L., Jain, B.,: Purlieus: locality-aware resource allocation for MapReduce in a cloud. In: Proceedings of the ACM/IEEE Conference on High Performance Computing Networking, Storage and Analysis, SC 2011, Seattle, WA, USA (2011)
22. Lu, L., Jin, H., Shi, X., Fedak, G.: Assessing MapReduce for internet computing: a comparison of Hadoop and BitDew-MapReduce. In: Proceedings of the 13th ACM/IEEE International

Benchmarking Virtualized Hadoop Clusters

Todor Ivanov[1,2(✉)], Roberto V. Zicari[1], and Alejandro Buchmann[2]

[1] Big Data Laboratory, Databases and Information Systems,
Goethe Universität Frankfurt am Main, Frankfurt, Germany
{todor, zicari}@dbis.cs.uni-frankfurt.de
[2] Databases and Distributed Systems, Technische Universität Darmstadt,
Darmstadt, Germany
buchmann@dvs.tu-darmstadt.de

Abstract. This work investigates the performance of Big Data applications in virtualized Hadoop environments, hosted on a single physical node. An evaluation and performance comparison of applications running on a virtualized Hadoop cluster with separated data and computation layers against standard Hadoop installation is presented. Our experiments show how different Data-Compute Hadoop cluster configurations, utilizing the same virtualized resources, can influence the performance of CPU bound and I/O bound workloads. Based on our observations, we identify three important factors that should be considered when configuring and provisioning virtualized Hadoop clusters.

Keywords: Big data · Benchmarking · Hadoop · Virtualization

1 Introduction

Apache Hadoop[1] has emerged as the predominant platform for Big Data applications. Recognizing this potential, Cloud providers have rapidly adopted it as part of their services (IaaS, PaaS and SaaS) [8]. For example, Amazon, with its Elastic MapReduce (EMR)[2] web service, has been one of the pioneers in offering Hadoop-as-a-service. The main advantages of such cloud services are quick automated deployment and cost-effective management of Hadoop clusters, realized through the pay-per-use model. All these features are made possible by virtualization technology, which is a basic building block of the majority of public and private Cloud infrastructures [7]. However, the benefits of virtualization come at a price of an additional performance overhead. In the case of virtualized Hadoop clusters, the challenges are not only the storage of large data sets, but also the data transfer during processing. Related works, comparing the performance of a virtualized Hadoop cluster with a physical one, reported virtualization overhead ranging between 2-10 % depending on the application type [1, 6, 9]. However, there were also cases where virtualized Hadoop performed better than the physical cluster, because of the better resource utilization achieved with virtualization.

---

[1] https://hadoop.apache.org/.
[2] http://aws.amazon.com/elasticmapreduce/.

© Springer International Publishing Switzerland 2015
T. Rabl et al. (Eds.): WBDB 2014, LNCS 8991, pp. 87–98, 2015.
DOI: 10.1007/978-3-319-20233-4_9

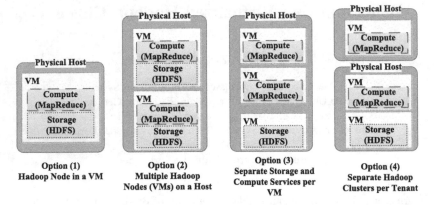

Fig. 1. Options for vrtualized Hadoop cluster deployments

In spite of the hypervisor overhead caused by Hadoop, there are multiple advantages of hosting Hadoop in a cloud environment [1, 6, 9] such as improved scalability, failure recovery, efficient resource utilization, multi-tenancy, security, to name a few. In addition, using a virtualization layer enables to separate the compute and storage layers of Hadoop on different virtual machines (VMs). Figure 1 depicts various combinations to deploy a Hadoop cluster on top of a hypervisor. Option (1) is hosting a worker node in a virtual machine running both a TaskTracker and NameNode service on a single host. Option (2) makes use of the multi-tenancy ability provided by the virtualization layer hosting two Hadoop worker nodes on the same physical server. Option (3) shows an example for functional separation of compute (MapReduce service) and storage (HDFS service) in separate VMs. In this case, the virtual cluster consists of two compute nodes and one storage node hosted on a single physical server. Finally, option (4) gives an example for two separate clusters running on different hosts. The first cluster consists of one data and one compute node. The second cluster consists of a compute node that accesses the data node of the first cluster. These deployment options are currently supported by Serengeti,[3] a project initiated by VMWare, and Sahara,[4] which is part of the OpenStack[5] cloud platform.

The objective of this work was to investigate the performance of Hadoop clusters, deployed with separated storage and compute layers (option (3)), on top of a hypervisor managing a single physical host. We have benchmarked different Hadoop cluster configurations by running CPU bound and I/O- bound workloads.

The rest of the paper is organized as follows: Sect. 2 presents the benchmarking approach; Sect. 3 describes the experimental platform; Sect. 4 presents the experiments and analyzes the results. Finally, Sect. 5 summarizes the lessons learned.

---

[3] http://www.projectserengeti.org/.

[4] https://wiki.openstack.org/wiki/Sahara/.

[5] http://www.openstack.org/.

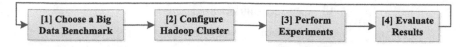

Fig. 2. Iterative experimental approach

2 An Iterative Approach to Big Data Benchmarking

We have defined and used what we call an *Iterative Experimental Approach* to investigate the performance of multiple test scenarios. The model is briefly illustrated in Fig. 2, and consists of four steps.

First step is to choose a Big Data benchmark which matches the characteristics of the evaluated application. For that, we have chosen the HiBench [2] benchmark suite. In each test scenario only one workload at a time is evaluated. In the second step, we configured the experimental platform by selecting the type of Hadoop cluster and allocating the virtualized resources to each cluster node. In the third step the experiments are performed following the rules of the chosen benchmark in order to ensure accurate and comparable test results. In the fourth and final step, the experimental results are analyzed by extracting the metrics, building graphical representations and validating the measured values.

We call our approach iterative because after the initial test run (completing steps 1 to 4) the user can start at step one, switching to a different workload, and continue performing new test runs on the same cluster configuration. However, to ensure consistent data state, a fresh copy of the input data has to be generated before each benchmark run. Similarly, in case of new cluster configuration all existing virtual nodes have to be deleted and replaced with new ones, using a basic virtual machine template. For all experiments, we ran exclusively only one cluster configuration at a time on the platform. In this way, we avoided biased results due to inconsistent system state.

3 Experimental Platform

An abstract view of the experimental platform we used to perform the tests is shown in Fig. 3. The platform is organized in four logical layers described below.

Hardware Layer: It consists of a standard Dell PowerEdge T420 server equipped with two Intel Xeon E5-2420 (1.9 GHz) CPUs each with six cores, 32 GB of RAM and four 1 TB, Western Digital (SATA, 3.5 in, 7.2 K RPM, 64 MB Cache) hard drives.

Management Layer (Virtualization): We installed the VMWare vSphere 5.1[6] platform on the physical server, including ESXi and vCenter Servers for automated VM management.

Platform Layer (Hadoop Cluster): Project Serengeti integrated in the vSphere Big Data Extension[7] (BDE) (version 1.0), installed in a separate VM, was used for automatic

---

[6] http://www.vmware.com/products/vsphere/.

[7] http://www.vmware.com/products/big-data-extensions/.

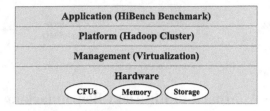

Fig. 3. Experimental platform layers

deployment and management of Hadoop clusters. The hard drives were deployed as separate datastores and used as shared storage resources by BDE. The deployment of both Standard and Data-Compute cluster configurations was done using the default BDE/Serengeti Server options as described in [4]. In all the experiments we used the Apache Hadoop distribution (version 1.2.1), included in the Serengeti Server VM template (hosting CentOS), with the default parameters: 200 MB java heap size, 64 MB HDFS block size and replication factor of 3.

Application Layer (HiBench): For our experiments, we have chosen two MapReduce representative applications from the HiBench micro-benchmarks, namely, the *Word-Count* (CPU bound) and the *TestDFSIOEnhanced* (I/O bound) workloads.

One obvious limitation of our experimental environment is that it consists of a single physical server, hosting all VMs, and does not involve any physical network communication between the VM nodes. Additionally, all experiments were performed on the VMWare ESXi hypervisor. This means that the reported results may not apply to other hypervisors as suggested by related work [3], comparing different hypervisors.

4　Experiments and Evaluation

4.1　Experimental Setup

The focus of our work is on analyzing the performance of different virtualized Hadoop cluster configurations deployed and tested on our platform. We identify two types of cluster configurations as depicted in Fig. 4: *Standard Hadoop (SH)* and *Data-Compute Hadoop (DCH)* clusters.

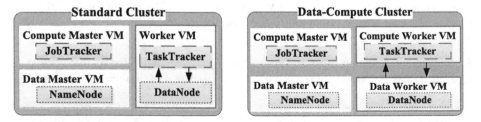

Fig. 4. Standard and Data-Compute Hadoop cluster configurations

The Standard Hadoop cluster type is a standard Hadoop cluster configuration but hosted in a virtualized environment with each cluster node installed in a separate VM. The cluster consists of one Compute Master VM (running *JobTracker*), one Data Master VM (running *NameNode*) and multiple Worker VMs. Each Worker VM is running both *TaskTracker (TT)* and *DataNode (DN)* services. *The data exchange is between the TT and DN services in the VM.*

On the other hand, the *Data-Compute Hadoop* cluster type has similarly Compute and Data Master VMs, but two types of Worker nodes: Compute and Data Worker VMs. This means that there are data nodes, running only DN service and compute nodes, running only TT service. *The data exchange is between the Compute and Data VMs, incurring extra network traffic.* The advantage of this configuration is that the number of data and compute nodes in a cluster can be independently and dynamically scaled, adapting to the workload requirements.

The first factor that we have to take into account when comparing the configurations is *the number of VMs utilized in a cluster.* Each additional VM increases the hypervisor overhead and therefore can influence the performance of a particular application as reported in [3, 9, 10]. At the same time, running more VMs utilizes more efficiently the hardware resources and in many cases leads to improved overall system performance (CPU and I/O Throughput) [9].

The second factor is that all cluster configurations should *utilize the same amount of hardware resources* in order to be comparable.

Taking these two factors into account, we specified six different cluster configurations. Two of the cluster configurations are of type *Standard Hadoop* cluster and the other four are of type *Data-Compute Hadoop* cluster. Based on the number of virtual nodes utilized in a cluster configuration, we compare *Standard1 with Data-Comp1* and *Standard2 with Data-Comp3* and *Data-Comp4*. Additionally, we added *Data-Comp2* to compare it with *Data-Comp1* and *Data-Comp3*. The goal is to better understand how the number of data nodes influences the performance of I/O bound applications in a *Data-Compute Hadoop* cluster.

Table 1 shows the worker nodes for each configuration and the allocated per VM resources (vCPUs, vRAM and vDisks). Three additional VMs (Compute Master, Data Master and Client VMs), not listed in Table 1, were used in all of the six cluster configurations. For simplicity, we will abbreviate in the rest, the *Worker Node as WN*, the *Compute Worker Node as CWN* and *Data Worker Node as DWN*.

4.2   CPU Bound Workload

In the first experiment, we investigated the performance of the *WordCount* [2] workload, which is a standard CPU bound MapReduce job. The workload takes three parameters: *input data size per node, number of map jobs per node and number of reduce jobs per node.* In the case of Data-Compute cluster these parameters are only relevant for the Compute Workers (running TaskTracker). Therefore, in order to achieve comparable results between the Standard and Data-Compute Hadoop cluster types, the *overall sum of the processed data and number of map and reduce tasks should be the same.* For example, to process 60 GB data in Standard1 (3 WNs) cluster

Table 1. Six experimental Hadoop cluster configurations

Configuration Name	Worker Nodes	
Standard1 (Standard Cluster 1)	**3 Worker Nodes**	
	TaskTracker & DataNode; 4 vCPUs; 4608 MB vRAM; 100 GB vDisk	
Standard2 (Standard Cluster 2)	**6 Worker Nodes**	
	TaskTracker & DataNode; 2 vCPUs; 2304 MB vRAM; 50 GB vDisk	
Data-Comp1 (Data-Compute Cluster 1)	**2 Compute Worker Nodes**	**1 Data Worker Node**
	TaskTracker; 5 vCPUs; 4608 MB vRAM; 50 GB vDisk	DataNode; 2 vCPUs; 4608 MB vRAM; 200 GB vDisk
Data-Comp2 (Data-Compute Cluster 2)	**2 Compute Worker Nodes**	**2 Data Worker Nodes**
	TaskTracker; 5 vCPUs; 4608 MB vRAM; 50 GB vDisk	DataNode; 1 vCPUs; 2084 MB vRAM; 100 GB vDisk
Data-Comp3 (Data-Compute Cluster 3)	**3 Compute Worker Nodes**	**3 Data Worker Nodes**
	TaskTracker; 3 vCPUs; 2664 MB vRAM; 20 GB vDisk	DataNode; 1 vCPUs; 1948 MB vRAM; 80 GB vDisk
Data-Comp4 (Data-Compute Cluster 4)	**5 Compute Worker Nodes**	**1 Data Worker Nodes**
	TaskTracker; 2 vCPUs; 2348 MB vRAM; 20 GB vDisk	DataNode; 2 vCPUs; 2048 MB vRAM; 200 GB vDisk

were configured 20 GB input data size, 4 map and 1 reduce tasks, whereas in Data-Comp1 (2 CWNs & 1 DWN) cluster were configured 30 GB input data size, 6 map and 1 reduce tasks. Similarly, we adjusted the input parameters for the remaining four clusters to ensure that the same amount of data was processed. We experimented with three different data sets (60, 120 and 180 GB), which compressed resulted in smaller sets (15.35, 30.7 and 46 GB).

Figure 5 depicts the WordCount completion times normalized for each input data size with respect to Standard1 as baseline. *The lower values represent faster completion times, respectively the higher values account for longer completion times.*

Table 2 compares cluster configurations utilizing the same number of VMs. In the first case, Standard1 (3 WNs) performs slightly (2 %) better than Data-Comp1 (2 CWNs & 1 DWN). In the second and third case, Standard2 (6 WNs) is around 23 % faster than Data-Comp3 (3 CWNs & 3 DWNs) and around 14 % faster than Data-Comp4 (5 CWNs & 1 DWNs), making it the best choice for CPU bound applications.

In Table 3, comparing the configurations with different number of VMs, we observe that Standard2 (6 WNs) is between 17-19 % faster than Standard1 (3 WNs), although Standard2 utilizes 6 VMs and Standard1 only 3 VMs. Similarly, Data-Comp4 (5 CWNs & 1 DWNs) achieves between 1-3 % faster times than Standard1 (3 WNs).

Table 2. WordCount comparison - equal number of VMs

Equal Number of VMs	3 VMs	6 VMs	6 VMs
Data Size (GB)	Diff. (%) Standard1/ Data-Comp1	Diff. (%) Standard2/ Data-Comp3	Diff. (%) Standard2/ Data-Comp4
60	0	22	13
120	2	22	14
180	2	23	14

Table 3. WordCount comparison - different number of VMs

Different Number of VMs	3 VMs	4 VMs
	4 VMs	6 VMs
Data Size (GB)	Diff. (%) Standard1/ Standard2	Diff. (%) Standard1/ Data-Comp4
60	-19	-3
120	-18	-1
180	-17	-1

In both cases having more VMs utilizes better the underlying hardware resources, which complies to the conclusions reported in [9].

Another interesting observation is that cluster Data-Comp1 (2 CWNs & 1 DWN) and Data-Comp2 (2 CWNs & 2 DWN) perform alike, although Data-Comp2 utilizes an additional instance of data worker node, which causes extra overhead on the hypervisor. However, as the *WordCount* workload is 100 % CPU bound, all the processing is performed on the compute worker nodes and the extra VM instance does not impact the actual performance. In the same time, if we compare the times on Fig. 5 of all four Data-Compute cluster configurations, we observe that the Data-Comp4 (5 CWNs & 1 DWNs) performs best. This shows both that the allocation of virtualized resources influence the application performance and that for CPU bound applications having more compute nodes is beneficial.

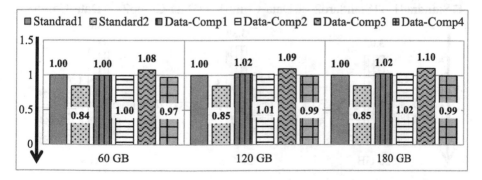

Fig. 5. Normalized WordCount completion times

Serengeti offers the ability for Compute Workers to use a Network File System (NFS) instead of virtual disk storage, also called TempFS in Serengeti. The goal is to ensure data locality, increase capacity and flexibility with minimal overhead. A detailed evaluation and experimental results of the approach are presented in the related work [5]. Using the TempFS storage type, we performed experiments with Data-Comp1 and Data-Comp4 cluster configurations. The results showed very slight 1-2 % improvement compared to the default shared virtual disk type that we used in all configurations.

4.3   I/O Bound Workload

In the next experiment, we use the *TestDFSIOEnhanced* [2] workload, which is designed to stress test the I/O storage (read and write) capabilities of a cluster. The workload consists of *TestDFSIO-read* and *TestDFSIO-write* parts, which will be discussed separately. The two test parameters used for all configurations were a constant file size of 100 MB and three variations of the number of files (100, 200 and 500), which resulted in three different data sets (10, 20 and 50 GB).

Figure 6 depicts the normalized *TestDFSIO-read* times, with Standard1 (3 WNs) achieving the best times for all test cases. Table 4 compares the cluster configurations utilizing the same number of VMs, whereas Table 5 compares configurations utilizing different number of VMs.

In the first case, Standard1 (3 WNs) performs up to 73 % better than Data-Comp1 (2 CWNs & 1 DWN) because of the different data placement strategies. In Data-Comp1 the data is stored on a single data node and should be read in parallel by the two compute nodes, which is not the case in Standard1 where each node stores the data locally, avoiding any communication conflicts. In the second case, Standard2 (6 WNs) performs between 18-46 % slower than Data-Comp3 (3 CWNs & 3 DWNs). Although, each node in Standard2 stores a local copy of the data, it seems that the resources allocated per VM are not sufficient to run both TT and DN services, which is not the case in Data-Comp3.

In Table 5 we observe that Data-Comp2 (2 CWNs & 2 DWNs) completion times are *two times* faster than Data-Comp1 (2 CWNs & 1 DWN). On the other hand, Data-Comp3, which utilizes 3 data nodes, is up to 39 % faster than Data-Comp2.

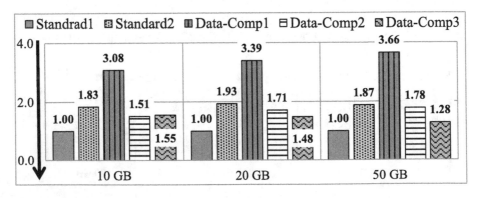

Fig. 6. Normalized TestDFSIOEnhanced read completion times

Table 4. TestDFSIOEnhanced read - equal number of VMs

Equal Number of VMs	3 VMs	6 VMs
Data Size (GB)	Diff. (%) Standard1/Data-Comp1	Diff. (%) Standard2/Data-Comp3
10	68	-18
20	71	-30
50	73	-46

Table 5. TestDFSIOEnhanced read - Different Number of VMs

Different Number of VMs	3 VMs	4 VMs
	4 VMs	6 VMs
Data Size (GB)	Diff. (%) Data-Comp1/Data-Comp2	Diff. (%) Data-Comp2/Data-Comp3
10	-104	3
20	-99	-15
50	-106	-39

This complies with our assumption that using more data nodes improves the read performance.

Figure 7 illustrates the *TestDFSIO-write* [2] completion times for five cluster configurations. Table 6 compares the cluster configurations utilizing the same number of VMs. In the first case, Standard1 (3 WNs) performs between 10-24 % slower than Data-Comp1 (2 CWNs & 1 DWN). The reason for this is that Data-Comp1 utilizes only one data node and the HDFS pipeline writing process writes all three block copies locally on the node, which of course is against the fault tolerance practices in Hadoop. In a similar way, Data-Comp3 (3 CWNs & 3 DWNs) achieves up to 14 % better times than Standard2 (6 WNs).

Table 7 compares cluster configurations with different number of VMs. Data-Comp1 (2 CWNs & 1 DWN) achieves up to 19 % better times than Data-Comp3

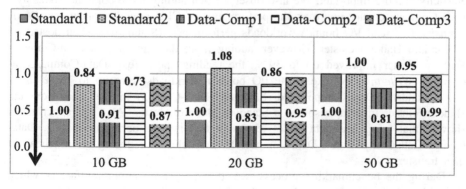

Fig. 7. Normalized TestDFSIOEnhanced write times

Table 6. TestDFSIOEnhanced Write - *Equal* Number of VMs

Equal Number of VMs	3 VMs	6 VMs
Data Size (GB)	Diff. (%) Standard1/Data-Comp1	Diff. (%) Standard2/Data-Comp3
10	-10	4
20	-21	-14
50	-24	-1

Table 7. TestDFSIOEnhanced Write - *Different* Number of VMs

Different Number of VMs	3 VMs	3 VMs
	6 VMs	6 VMs
Data Size (GB)	Diff. (%) Data-Comp1/Data-Comp3	Diff. (%) Standard1/Data-Comp3
10	-4	-15
20	13	-6
50	19	-1

(3 CWNs & 3 DWNs), because of the extra cost of writing to 3 data nodes (enough to guarantee the minimum data fault tolerance) instead of only one data node. Further observations show that although Data-Comp3 utilizes 6VMs, it achieves up to 15 % better times than Standard1, which utilizes only 3VMs. However, this difference decreases from 15 % to 1 % with the growing data sizes and may completely vanish for larger data sets.

5   Lessons Learned and Future Work

Our experiments showed that *compute-intensive* (i.e. CPU bound) workloads are more suitable for Standard Hadoop clusters, but the overhead (up to 14 %) when hosted on a Data-Compute Hadoop cluster is acceptable in cases when node elasticity is more important. We also observed that adding more compute nodes to a Data-Compute cluster improves the performance of CPU bound applications. *Read-intensive* (i.e. read I/O bound) workloads perform best (Standard1) when hosted on a Standard Hadoop cluster. However, adding more data nodes to a Data-Compute Hadoop cluster improved up to 39 % the reading speed (e.g. Data-Comp2/Data-Comp3). *Write-intensive* (i.e. write I/O bound) workloads were up to 15 % faster (e.g. Standard2/Data-Comp3 and Standard1/Data-Comp3) on a Data-Compute Hadoop cluster in comparison to a Standard Hadoop cluster. Also our experiments showed that using less data nodes results in better write performance (e.g. Data-Comp1/Data-Comp3) on a Data-Compute Hadoop cluster, reducing the overhead of data transfer.

During the benchmarking process, we identified three important factors which should be taken into account when configuring a virtualized Hadoop cluster:

1. *Choosing the "right" cluster type (Standard or Data-Compute Hadoop cluster) that provides the best performance for the hosted Big Data workload is not a straightforward process.* It requires very precise knowledge about the workload type, i.e. whether it is CPU intensive, I/O intensive or mixed, as indicated in Sect. 4.
2. *Determining the number of nodes for each node type (compute and data nodes) in a Data-Compute cluster is crucial for the performance and depends on the specific workload characteristics.* The extra network overhead, caused by intensive data transfer between data and compute worker nodes, should be carefully considered, as reported by Ye et al. [10].
3. *The overall number of virtual nodes running in a cluster configuration has direct influence on the workload performance,* this is also confirmed by [3, 9, 10]. Therefore, it is crucial to choose the optimal number of virtual nodes in a cluster, as each additional VM causes an extra overhead to the hypervisor. At the same time, we observed cases, e.g. *Standard1/Standard2* and *Data-Comp1/Data-Comp2,* where clusters consisting of more VMs utilized better the underlying hardware resources.

In the near future, we plan to repeat the experiments on a virtualized multi-node cluster involving physical network overhead. Additionally, we will also consider experimenting with other workloads (e.g. mixed CPU and I/O bound) as well as using larger data sets. We are also planning to perform similar experiments using OpenStack and other hypervisors (e.g. KVM and Xen).

Acknowledgments. We would like to thank Jeffrey Buell of VMware for providing a useful feedback on an early version of this paper and Nikolaos Korfiatis for his helpful comments and support.

References

1. Buell, J.: Virtualized Hadoop Performance with VMware vSphere 5.1. Tech. White Pap. VMware Inc. (2013)
2. Huang, S., et al.: The HiBench benchmark suite: characterization of the MapReduce-based data analysis. In: 2010 IEEE 26th International Conference on Data Engineering Workshops (ICDEW), pp. 41–51. IEEE (2010)
3. Li, J., et al.: Performance overhead among three hypervisors: an experimental study using hadoop benchmarks. In: 2013 IEEE International Congress on Big Data (BigData Congress), pp. 9–16. IEEE (2013)
4. Li, X., Murray, J.: Deploying Virtualized Hadoop Systems with VMWare vSphere Big Data Extensions. Tech. White Pap. VMware Inc. (2014)
5. Magdon-Ismail, T., et al.: Toward an elastic elephant enabling hadoop for the Cloud. VMware Tech. J. (2013)
6. Microsoft: Performance of Hadoop on Windows in Hyper-V Environments. Tech. White Pap. Microsoft (2013)
7. Rimal, B.P., et al.: A taxonomy and survey of cloud computing systems. In: Fifth International Joint Conference on INC, IMS and IDC, NCM 2009, pp. 44–51. Ieee (2009)

8. Schmidt, R., Mohring, M.: Strategic alignment of cloud-based architectures for big data. In: 2013 17th IEEE International Enterprise Distributed Object Computing Conference Workshops (EDOCW), pp. 136–143 IEEE (2013)
9. VMWare: Virtualized Hadoop Performance with VMware vSphere ®5.1. Tech. White Pap. VMware Inc. (2013)
10. Ye, K., et al.: vHadoop: a scalable hadoop virtual cluster platform for MapReduce-based parallel machine learning with performance consideration. In: 2012 IEEE International Conference on Cluster Computing Workshops (Cluster Workshops), pp. 152–160. IEEE (2012)

In-Memory, Data Generation
and Graphs

PopulAid: In-Memory Test Data Generation

Ralf Teusner[1]([⊠]), Michael Perscheid[2], Malte Appeltauer[2], Jonas Enderlein[1],
Thomas Klingbeil[2], and Michael Kusber[2]

[1] Hasso Plattner Institute, University of Potsdam, Potsdam, Germany
`ralf.teusner@hpi.uni-potsdam.de`,
`jonas.enderlein@student.hpi.uni-potsdam.de`
[2] SAP Innovation Center, Potsdam, Germany
`{Michael.Perscheid,Malte.Appeltauer,Thomas.Klingbeil,`
`Michael.Kusber}@sap.com`

Abstract. During software development, it is often necessary to access real customer data in order to validate requirements and performance thoroughly. However, company and legal policies often restrict access to such sensitive information. Without real data, developers have to either create their own customized test data manually or rely on standardized benchmarks. While the first tends to lack scalability and edge cases, the latter solves these issues but cannot reflect the productive data distributions of a company.

In this paper, we propose *PopulAid* as a tool that allows developers to create customized benchmarks. We offer a convenient data generator that incorporates specific characteristics of real-world applications to generate synthetic data. So, companies have no need to reveal sensible data but yet developers have access to important development artifacts. We demonstrate our approach by generating a customized test set with medical information for developing SAP's healthcare solution.

Keywords: In-memory database · Data generation · Application testing

1 Introduction

Development and maintenance of enterprise applications highly depends on productive data to take the risk of slow, wrong, or inconsistent information into account. These issues can be handled best with productive data because it embodies the optimal basis for adding new features, debugging, and profiling performance bottlenecks. So, developers can ensure that their enterprise applications are able to handle expected data and increased usage in the future.

However, in most cases it is impossible to develop with productive data. Legal reasons, competitive advantages, clues to business secrets, and data privacy usually prohibit the usage of such data. Unfortunately, developers need test data, therefore they have to spend parts of their time on the creation of data that suits their needs. For two reasons, this data is likely to be biased in some way:

© Springer International Publishing Switzerland 2015
T. Rabl et al. (Eds.): WBDB 2014, LNCS 8991, pp. 101–108, 2015.
DOI: 10.1007/978-3-319-20233-4_10

– Developers cannot know every single detail of productive data so that they tend to miss important requirements and edge cases. In consequence, the generated data is based on false assumptions and does not reflect productive data in all circumstances.
– Time is limited. Therefore, the creation is done as simple as possible and the amount of generated data is limited to the amount absolutely required–scalability is not tested under these conditions.

To solve these issues, standardized benchmarks are a means to help developers during development of large applications. They offer not only realistic data volumes but also cover possible edge cases and faulty inputs. However, based on the fact that they are standardized, they can neither reflect the productive data distributions of a certain company nor include individual queries sufficiently [1].

To circumvent these shortcomings, an ideal solution would be customized benchmarks with *generated test data* that shares the same characteristics as in real applications. While the standard benchmark includes appropriate and common edge cases, the customization reflects specific tables, productive data volumes, and distributions present in a company. This way, developers would get a scalable dataset to test specific queries without requiring customers to reveal their sensitive information.

In this paper, we present PopulAid[1] as a tool for generating customized test data. With the help of our web interface, developers can easily adjust their schemas, assign generators to columns, and get immediate previews of potential results. In doing so, generators configure not only specific value properties for one column such as data type, range and data pools, distribution, or number of distinct and undefined values; but also dependencies for column combinations such as foreign keys, pattern evaluation, and functional relations. PopulAid allows developers to create data in a scalable and efficient manner by applying these generators to SAP HANA. This columnar in-memory database can leverage the performance potentials also for write-intensive tasks such as data generation [2]. We present our approach with the help of a real-world example that generates a test set representing medical data from SAP's healthcare solution.

The remainder of this paper is structured as follows: Sect. 2 presents PopulAid and its core features, Sect. 3 evaluates our approach, Sect. 4 discusses related work, and Sect. 5 concludes.

2 PopulAid

PopulAid is designed to be a convenient solution for realistic application testing. We seamlessly integrate our data generation into development processes and offer different ways of interacting with our approach in order to support good usability. To reach this goal, we focus on three core concepts:

---

[1] More information (including a screencast) can be found at: https://epic.hpi.uni-potsdam.de/Home/PopulAid.

– No setup is required. When available, PopulAid is shipped together with the target database SAP HANA.
– Immediate feedback of the input via a preview of the values to be generated. As depicted in Fig. 1, for every column, a representative selection of ten values is shown directly in the web front-end.
– Assistive guessing of suitable generators. Especially for wide tables with more than 100 columns, assigning generators manually is tedious. Guessing of suitable generators for missing columns is done on the basis of the present datatype, the name of the column, and past generation tasks. Columns that have a not null constraint and did not get assigned a generator manually, are chosen automatically.

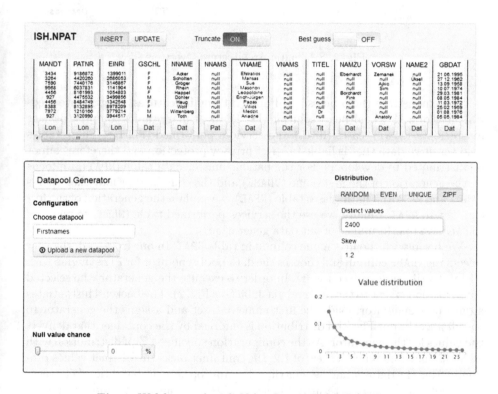

Fig. 1. Webfrontend with Value Previews for Table NPAT

The main interface is a web front-end as shown in Fig. 1. Additionally, data generation can also be configured via an Eclipse plugin or directly via a Java API, which, for example, allows unit tests to automatically generate data.

Even if PopulAid was initially aimed for SAP HANA, its API supports arbitrary databases reachable via JDBC. Concerning the user interface, a generic approach to retrieve existing distributions and present immediate feedback is currently under development.

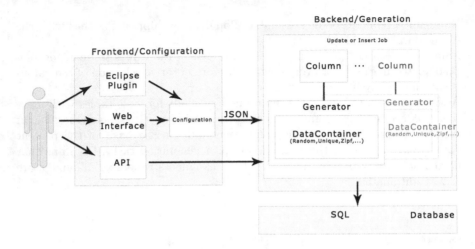

Fig. 2. Schematic Architecture Overview

To demonstrate the process and PopulAid's main features, we generate medical data for SAP's healthcare solution (IS-H). Its primary tables comprise patient master data (NPAT), cases (NFAL) and services performed (NLEI)[2]. These tables hold sensitive data that is liable to strict privacy protection and therefore cannot be distributed to developers. For the patient master data table (NPAT), we focus on the generation of the first name (VNAME) and the combined primary key (fields MANDT and PATNR). On the cases table (NFAL), we explain the generation of the foreign key to NPAT. Finally, we use the services performed table (NLEI) to measure the achieved performance of our data generation.

We begin with the first name column in table NPAT. In our front-end, the user selects the VNAME column and chooses the data pool generator for Firstnames with a Zipf distribution [3] (see Fig. 1). In order to execute the generator, the selected configuration is sent to the backend via JSON (see Fig. 2). The backend instantiates a data pool generator, loads the first name dataset and assigns the generator to the VNAME column. The Zipf distribution is enforced by the container that delivers the values for the generator. As the configuration specifies a Zipf distribution with 2400 distinct values and a skew of 1.2, the container picks the distinct values randomly from the dataset and assigns them the appropriate probabilities. Whenever the generator is called, it returns its pick from the containers' weighted value set. After all generators of a table were called, the picks are set as values for a prepared insert statement, which is then executed in batches.

For the combined primary key in NPAT, we create long generators for both columns (MANDT and PATNR). To enforce uniqueness of the combinations, the generators are ordered and handed over to an instance of MultiColumnGenerator

---

[2] More information concerning the tables and the included attributes is accessible under: http://help.sap.com/saphelp_crm60/helpdata/de/09/a4d2f5270f4e58b2358fc5519283be/content.htm.

with distribution set to unique. Afterwards, the `MultiColumnGenerator` is assigned to both columns. On each call, the generator returns the next entry from an unique array of values for the columns it is assigned to.

In table `NFAL`, the fields `MANDT` and `PATNR` realize the connection of the case to the patient master data. Therefore, we use a `FunctionalDependencyGenerator` here. This generator retrieves all distinct values from the source columns `PATNR` and uses them as its value pool. Apart from that, the `ForeignKeyGenerator` works like the data pool generator.

In addition to the aforementioned generators, PopulAid features various other generators to create dates and timestamps, texts that follow a specified pattern, e.g., for phone numbers or email addresses, or use custom SQL queries to extract their value pools. With regard to data types, all standard types are supported. To reflect realistic data, it is also possible to pollute the generated values with a predefined share of undefined or null values. PopulAid also allows to update values in existing tables, for example to add missing columns, apply expected characteristics to already existing datasets, complete sparse data or to provoke certain edge cases.

3   Evaluation

After creating data with PopulAid, we have a more detailed look into its performance characteristics. For that reason, we generate data for our previous introduced tables (`NPAT`: 116 columns, `NFAL`: 76 columns, `NLEI`: 112 columns). These tables also have specific requirements that have to be fulfilled: `NPAT` has a two multi-column uniqueness constraint over two columns; `NFAL` has a random foreign key dependency over two combined columns to `NPAT`; and `NLEI` has a random foreign key dependency over two combined columns to `NFAL`.

We measure the performance for creating our customized test data under the following conditions. We consider both the number of entries per second and the raw data throughput (in MB per second).

When profiling the generator, we focus on the pure execution time and exclude the setup time that has to be done only once. We choose a batch size of 20 entries to be inserted together, which turned out to be optimal in most cases. The inserting transaction is always committed just at the very end of the generation. To rule out statistical outliers, we perform each measurement 10 times and choose the median value. We execute all benchmarks on an Intel Core i7 4 x 2.6 GHz CPU with 4 GB RAM assigned to the Java Virtual Machine, connected via Ethernet to a virtual machine-based SAP HANA running on a Xeon X5670 with 4 cores at 2.93 GHz with 32 GB RAM.

Table 1 presents our performance measurements for SAP's healthcare solution. As can be seen in the table, the run-time of PopulAid's generation process increases nearly linearly when incrementing the data size. For example, generating data for `NFAL` requires 2,5 s for 100,000 entries and 25 s for 1,000,000 entries. Figure 3 also illustrates the linear correlation between generated entries and the runtime. The throughput seems to be constantly high with generated entries

Table 1. Performance Measurements for NPAT, NFAL, and NLEI Tables

	Run-time in ms			Entries/s			MB/s		
Data Size	NPAT	NFAL	NLEI	NPAT	NFAL	NLEI	NPAT	NFAL	NLEI
1,000	120	51	60	8,333	19,608	16,667	9.63	13.84	17.01
10,000	387	296	360	25,839	33,783	27,777	29.83	23.85	28.33
100,000	3,844	2,505	3,305	26,014	39,920	30,257	30.11	28.21	30.86
1,000,000	45,611	25,550	34,112	21,925	39,139	29,315	25.37	27.66	29.90

between 20,000 and 40,000 per second. Only for small amounts of data size, the throughput is considerably lower than the throughputs achieved for larger amounts. This can be explained with the higher influence of opening transactions and accompanying database connections on the entire performance. For greater datasets than shown in Table 1, the throughput tested on NLEI remained constant at about 30,000 entries per second for 5 million entries. The differences in throughput between tables is based on the individual constraints for data generation. A probable reason for the decrease in NPAT is the uniqueness requirement that has to be fulfilled. Considering NLEI, the greater size in terms of columns may be the reason for lower entries per second in comparison to NFAL.

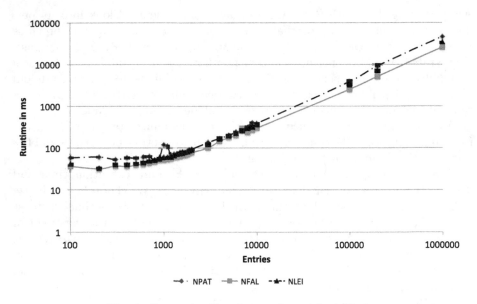

Fig. 3. Generation Runtime on Logarithmic Scale

While conducting our performance tests, we spotted the JDBC connection as the limiting factor of throughput. For that reason, we still wanted to measure the theoretical throughput without network limitations. When writing the data to disk into a comma-separated values (CSV) file, for each table the achieved throughput ranged between 29,9 MB/s and 33,4 MB/s. This speed is close to the optimal performance when writing to the database because the CPU utilizations while inserting was fluctuating in the 90 % range. Nevertheless, we can still achieve higher writing throughput if we implement full parallelization in the near future.

4   Related Work

In general, data generation is a widely researched area in computer science, which would go beyond the scope of this paper. In the context, we therefore focus on related work concerning data generation in the context of customized benchmarks for application testing. The need for customizing standard benchmarks such as defined by the Transaction Processing Council (TPC) has already been identified [1]. The authors figured out that the domain-specific benchmarks are increasingly irrelevant due the diversity of data-centric applications. For that reason, they call for techniques and tools to synthetically scaling data and create application-specific benchmarks. The multi-dimensional data generator (MUDD) [4] allows for generating huge data volumes with appropriate distributions and hierarchies. However, it is designed specifically to support the tables used in the TPC-DS benchmark. To generate data for customized tables, their configuration has to be implemented as ANSI-C classes. The parallel data generation framework (PDGF) [5] supports the processing, storage, and loading aspects of big data analytics and so allows for end-to-end benchmarks. This generic data generator is currently applied in TPC benchmarks because it generates large amount of relational data very fast by parallelizing with seeded random number generators. The big data generator suite (BDGS) [6] reuses PDGF and enhances the approach with the 4 V requirements (volume, variety, velocity, and veracity). For example, it derives characteristics from real data and so preserves data veracity or supports several data sources and types in order to achieve variety. Finally, Myriad [7] is an expressive data generation toolkit that makes extensive use of parallelization.

Compared to PopulAid, the flexibility of the presented approaches for customized application test data is limited. While the aforementioned projects cover various aspects of the 4 V requirements, they lack capabilities to configure the generation in a fast and intuitive way. In order to play out their technical features, generators have to be easily usable and fit into the development process seamlessly. PopulAid satisfies this requirement with an intuitive web front-end, immediate feedback about expected results, and semi-automatic configuration of generators.

5   Conclusion

In this paper, we presented a data generation tool named PopulAid. It allows developers to easily create customized test data when productive data is not available. For this purpose, PopulAid offers a convenient web interface in order to adjust schemas, get assisted with generators, and obtain immediate previews. Our approach offers a broad spectrum to generate data from adapting specific values to defining complex dependencies between multiple columns. We integrated PopulAid into the SAP HANA in-memory database and showed how to generate a customized test data set for medical data from SAP's healthcare solution.

Future work deals with two topics. First, we will further improve the performance of PopulAid. Currently, we are working on a fully parallelized solution that is directly integrated into the SAP HANA in-memory database.

Second, we are experimenting with different approaches concerning the anonymization of customer data. Instead of the "traditional" way, generating data with similar characteristics, we experiment with means to manipulate real-world data until a back reference is not possible anymore. This method allows developers to create customized benchmarks which are still closer to customer data.

Acknowledgments. We thank Janusch Jacoby, Benjamin Reissaus, Kai-Adrian Rollmann, and Hendrik Folkerts for their valuable contributions during the development of PopulAid.

References

1. Tay, Y.C.: Data generation for application-specific benchmarking. In: VLDB Challenges and Visions, pp. 1470–1473 (2011)
2. Plattner, H.: A Course in In-Memory Data Management. Springer, Heidelberg (2013)
3. Newman, M.E.: Power laws, pareto distributions and zipf's law. Contemp. Phys. **46**(5), 323–351 (2005)
4. Stephens, J.M., Poess, M.: MUDD: a multi-dimensional data generator. In: Proceedings of the 4th International Workshop on Software and Performance. WOSP 2004, pp. 104–109, ACM (2004)
5. Rabl, T., Jacobsen, H.-A.: Big data generation. In: Rabl, T., Poess, M., Baru, C., Jacobsen, H.-A. (eds.) WBDB 2012. LNCS, vol. 8163, pp. 20–27. Springer, Heidelberg (2014)
6. Ming, Z., Luo, C., Gao, W., Han, R., Yang, Q., Wang, L., Zhan, J.: BDGS: A scalable big data generator suite in big data benchmarking, pp. 1–16 (2014). arXiv preprint arXiv:1401.5465
7. Alexandrov, A., Tzoumas, K., Markl, V.: Myriad: scalable and expressive data generation. Proc. VLDB Endow. **5**(12), 1890–1893 (2012)

Towards Benchmarking IaaS and PaaS Clouds for Graph Analytics

Alexandru Iosup[1]([✉]), Mihai Capotă[1], Tim Hegeman[1], Yong Guo[1], Wing Lung Ngai[1], Ana Lucia Varbanescu[2], and Merijn Verstraaten[2]

[1] Delft University of Technology, Delft, The Netherlands
A.Iosup@tudelft.nl
[2] University of Amsterdam, Amsterdam, The Netherlands
A.L.Varbanescu@uva.nl

Abstract. Cloud computing is a new paradigm for using ICT services— only when needed and for as long as needed, and paying only for service actually consumed. Benchmarking the increasingly many cloud services is crucial for market growth and perceived fairness, and for service design and tuning. In this work, we propose a generic architecture for benchmarking cloud services. Motivated by recent demand for data-intensive ICT services, and in particular by processing of large graphs, we adapt the generic architecture to Graphalytics, a benchmark for distributed and GPU-based graph analytics platforms. Graphalytics focuses on the dependence of performance on the input dataset, on the analytics algorithm, and on the provisioned infrastructure. The benchmark provides components for platform configuration, deployment, and monitoring, and has been tested for a variety of platforms. We also propose a new challenge for the process of benchmarking data-intensive services, namely the inclusion of the data-processing algorithm in the system under test; this increases significantly the relevance of benchmarking results, albeit, at the cost of increased benchmarking duration.

1 Introduction

Cloud services are an important branch of commercial ICT services. Cloud users can provision from Infrastructure-as-a-Service (IaaS) clouds "processing, storage, networks, and other fundamental resources" [43] and from Platform-as-a-Service (PaaS) clouds "programming languages, libraries, [programmable] services, and tools supported by the provider" [43]. These services are provisioned on-demand, that is, when needed, and used for as long as needed and paid only to the extent to which they are actually used. For the past five years, commercial cloud services provided by Amazon, Microsoft, Google, etc., have gained an increasing user base, which includes small and medium businesses [5], scientific HPC users [16,35], and many others. Convenient and in some cases cheap cloud services have enabled many new ICT applications. As the market is growing and diversifying, benchmarking and comparing cloud services, especially from commercial cloud providers, is becoming increasingly more important. In this

© Springer International Publishing Switzerland 2015
T. Rabl et al. (Eds.): WBDB 2014, LNCS 8991, pp. 109–131, 2015.
DOI: 10.1007/978-3-319-20233-4_11

work, we study the process of benchmarking IaaS and PaaS clouds and, among the services provided by such clouds for new ICT applications, we focus on graph analytics, that is, the processing of large amounts of linked data.

Benchmarking is a traditional approach to verify that the performance of a system meets the requirements. When benchmarking results are published, for example through mixed consumer-provider organizations such as SPEC and TPC, the consumers can easily compare products and put pressure on the providers to use best-practices and perhaps lower costs. For clouds, the new use for benchmarking results is to convince customers about the performance, the elasticity, the stability, and the resilience of offered services, and thus to convince customers to rely on cloud services for the operation of their businesses.

Because of its many uses, benchmarking has been the focus of decades of scientific and practical studies. There are many successful efforts on benchmarking middleware [7,8], on benchmarking databases [24], on the performance evaluation of grid and parallel-system schedulers [12,17,20,31], and on benchmarking systems in general [4,36].

To benchmark IaaS and PaaS clouds, older benchmarking techniques need to be adapted and extended. As an example of adaptation, cloud benchmarks need to adapt traditional techniques to the new cloud-workloads. We conjecture that the probable characteristics of current and near-future workloads can be derived from three major trends emerging from the last decade of grid and large-scale computing. First, individual jobs are now predominantly split into smaller compute or data-intensive tasks (many tasks [51]); there are almost no tightly coupled parallel jobs. Second, the duration of individual tasks is diminishing with every year; few tasks are still running for longer than one hour and a majority require only a few minutes to complete. Third, compute-intensive jobs are split either into bags-of-tasks (BoTs) or DAG-based workflows, but data-intensive jobs may use a variety of programming models, from MapReduce to general dataflow.

As an example of extension, cloud benchmarks need to extend traditional techniques to accommodate the new application domains supported by clouds. Currently, the use of clouds is fragmented across many different application areas, including hosting applications, media, games, and web sites, E-commerce, On-Demand Workforce and CRM, high-performance computing, search, and raw resources for various usage. Each application area has its own (de facto) performance standards that have to be met by commercial clouds, and some have even developed benchmarks (e.g., BioBench [3] for Bioinformatics and RUBiS [57] for online business). More importantly, many of these applications have rely upon unique, deep, and distributed software stacks, which pose many unresolved challenges to traditional benchmarking approaches—even the definition of the system under test becomes complicated.

We discuss in this article a generic approach to IaaS and PaaS cloud benchmarking. We propose a generic architecture for IaaS and PaaS cloud benchmarking. We have designed the architecture so that it is already familiar to existing practitioners, yet provide new, cloud-specific functionality. For example, current IaaS cloud operators lease resources to their customers, but leave the selection of resource types and the selection of the lease/release moments as a customer task; because such selection can impact significantly the performance of the system built to use the leased resources, the generic benchmarking architecture must include policies for provisioning and allocation of resources. Similarly, the current platforms may require the configuration of deep stacks of software (middleware), so the generic benchmarking architecture much include policies for advanced platform configuration and management.

Although we have designed the architecture to be generic, we have not yet proven that it can be used to benchmark the vast diversity of existing cloud usage scenarios. Authors of this article have already used it, in practice, to benchmark a variety of IaaS cloud usage scenarios [32,59,61]. Motivated by the increasingly important fraction of application data, in some cases over three-quarters (IHS and Cisco studies in April 2014) or even higher (IDC Health Insights report in December 2014), that is already or will soon reside in the cloud, we propose in this work an application of the generic architecture for *data-intensive* applications, in the graph-processing application domain. We focus on *graph analytics*, which is a data-intensive process that is increasingly used to provide insight into social networks, personalized healthcare, natural language processing, and many other fields of science, engineering, retail, and e-government. As a consequence, several well established graph analytics platforms, such as GraphLab, Giraph, and GraphX, are competing with many graph analytics platforms that are currently in operation or under development. Due to the high diversity of workloads encountered in practice, both in terms of algorithm and of dataset, and to the influence of these workloads on performance, it is increasingly difficult for users to identify the graph analytics platform best suited for their needs. We propose in this work Graphalytics, a benchmarking approach that focuses on graph analytics and is derived from the generic architecture. Graphalytics focuses on understanding how the performance of graph analytics platforms depends on the input dataset, on the analytics algorithm, and on the provisioned infrastructure. It provides components for platform configuration, deployment, and monitoring. It also manages the benchmarking process, from configuration to reporting.

Last, starting from the experience we have accumulated designing and using Graphalytics in practice, we identify an important new challenge for benchmarking in clouds: the need to understand the platform-specific and data-dependent performance of algorithms used for graph analytics. For clouds, it is currently not possible for customers to have access to the low-level technical specifications of the service design, deloyment, and tuning, and thus also not possible to easily select a particular algorithm to implement and deploy on the provisioned platform. Instead, either the cloud customers or the cloud operators need to determine an appropriate algorithm, for example through benchmarking. Thus, for clouds we see the need to include the algorithm in the system under test. We show preliminary results that give strong evidence for this need, and discuss steps towards algorithm-aware benchmarking processes.

This article has evolved from several regular articles [19,25,27], a book chapter [33], and a series of invited talks given by the authors between

2012 and 2014, including talks at MTAGS 2012 [34], HotTopiCS 2013 [29], etc[1]. This work has also benefited from valuable discussion in the SPEC Research Group's Cloud Working Group and Big Data Working Group. The new material in this article focuses on the application of the general architecture for IaaS cloud benchmarking to graph analytics, and on a new challenge for benchmarking big data processes, related to the inclusion of processing-algorithm alternatives in the evaluation process.

The remainder of this article is structured as follows. In Sect. 2, we present a primer on benchmarking computer systems. Then, we introduce a generic approach for IaaS cloud benchmarking, in Sect. 3. We apply the generic architecture to graph analytics, and propose the resulting benchmarking framework Graphalytics, in Sect. 4. We introduce a new challenge for cloud benchmarking, related to the inclusion of the algorithm in the system under test, in Sect. 5. Last, we conclude in Sect. 6.

2 A Primer on Benchmarking Computer Systems

We review in this section the main elements of the typical benchmarking process, which are basically unchanged since the early 1990s. For more detail, we refer to canonical texts on benchmarking [24] and performance evaluation [36] of computer systems. We also discuss the main reasons for benchmarking.

2.1 Elements of Benchmarking

Inspired by canonical texts [24,36], we review here the main elements of a benchmarking process. The main requirements of a benchmark—relevance, portability, scalability, and simplicity—have been discussed extensively in related literature, for example in [24, Chap. 1].

The *System Under Test (SUT)* is the system that is being evaluated. A *white box* system exposes its full operation, whereas a *black box* system does not expose operational details and is evaluated only through its outputs.

The *workload* is the operational load to which the SUT is subjected. Starting from the empirical observation that "20 % of the code consumes 80 % of the resources", simple *microbenchmarks (kernel benchmarks* [24, Chap. 9]) are simplified or reduced-size codes designed to stress potential system bottlenecks. Using the methodology of Saavedra et al. [53] and later refinements such as Sharkawi et al. [55], the results of microbenchmarks can be combined with application profiles to provide credible performance predictions for any platform. *Synthetic*

---

[1] In inverse chronological order: Lecture at the Fifth Workshop on Big Data Benchmarking (WBDB), Potsdam, Germany, August 2014. Lecture at the Linked Data Benchmark Council's Fourth TUC Meeting 2014, Amsterdam, May 2014. Lecture at Intel, Haifa, Israel, June 2013. Lecture at IBM Research Labs, Haifa, Israel, May 2013. Lecture at IBM T.J. Watson, Yorktown Heights, NY, USA, May 2013. Lecture at Technion, Haifa, Israel, May 2013. Online lecture for the SPEC Research Group, 2012.

and even *real-world (complex) applications* are also used for benchmarking purposes, as a response to system improvements that make microbenchmarks run fast but do not affect the performance of much larger codes. For distributed and large-scale systems such as IaaS clouds, *simple workloads* comprised of a single application and a (realistic) job arrival process represent better the typical system load and have been used for benchmarking [30]. *Complex workloads*, that is, the combined simple workloads of multiple users, possibly with different applications and job characteristics, have started to be used in the evaluation of distributed systems [30,59]; we see an important role for them in benchmarking.

The *Benchmarking Process* consists of the set of rules, prior knowledge (invariants), and procedures used to subject the SUT to the benchmark workload, and to collect and report the results.

2.2   Why Benchmarking?

Benchmarking computer systems is the process of evaluating their performance and other non-functional characteristics with the purpose of comparing them with other systems or with industry-agreed standards. Traditionally, the main use of benchmarking has been to facilitate the informed procurement of computer systems through the publication of verifiable results by system vendors and third-parties. However, benchmarking has grown as a support process for several other situations, which we review in the following.

Use in System Design, Tuning, and Operation: Benchmarking has been shown to increase pressure on vendors to design better systems, as has been for example the experience of the TPC-D benchmark [24, Chap. 3, Sect. IV]. For this benchmark, insisting on the use of SQL has driven the wide acceptance of the ANSI SQL-92; furthermore, the complexity of a majority of the queries has lead to the stress of various system bottlenecks, and ultimately to numerous improvements in the design of aggregate functions and support for them. This benchmark also led to a wide adoption of the geometric mean for aggregating normalized results [4]. The tuning of the DAS multi-cluster system has benefited from the benchmarking activity of some of the authors of this chapter, developed in the mid-2000s [30]; then, our distributed computing benchmarks exposed various (fixable) problems of the in-operation system.

Use in System Procurement: Benchmarks such as HPCC, the SPEC suites, the TPC benchmarks, etc. have long been used in system procurement—the systems tested with these benchmarks can be readily compared, so a procurement decision can be informed. Benchmarks can be also very useful tools for system procurement, even when the customer's workloads are not ideally matched by the workloads represented in the benchmark, or when the representativeness of the benchmark for an application domain can be questioned. In such situation, the customer gains trust in the operation of the system, rather than focus on the actual results.

Use in Training: One of the important impediments in the adoption of a new technology is the lack of expertise of potential users. Market shortages of qualified personnel in computer science are a major cause of concern for the European

Union and the US. Benchmarks, through their open-source nature and representation of industry-accepted standards, can represent best-practices and thus be valuable training material.

On Alternatives to Benchmarking: Several alternative methods have been used for the purposes described earlier in this section, among them empirical performance evaluation, simulation, and even mathematical analysis. We view benchmarking as an empirical evaluation of performance that follows a set of accepted procedures and best-practices. Thus, the use of empirical performance evaluation is valuable, but perhaps without the representativeness of a (de facto) standard benchmark. We see a role for (statistical) simulation [18,21,47] and mathematical analysis when the behavior of the system is well-understood and for long-running evaluations that would be impractical otherwise. However, simulating new technology, such as cloud computing, requires careful (and time-consuming) validation of assumptions and models.

3   A Generic Architecture for IaaS and PaaS Cloud Benchmarking

We propose in this section a generic architecture for IaaS and PaaS cloud benchmarking. Our architecture focuses on conducting benchmarks as sets of (real-world) experiments that lead to results with high statistical confidence, on considering and evaluating IaaS clouds as evolving black-box systems, on employing complex workloads that represent multi-tenancy scenarios, on domain-specific scenarios, and on a combination of traditional and cloud-specific metrics.

3.1   Overview

Our main design principle is to adapt the proven designs for benchmarking to IaaS clouds. Thus, we design an architecture starting from our generic architecture for IaaS cloud benchmarking [29,33,34], which in turn builds on our GrenchMark framework for grid benchmarking [30,61]. The result, a generic

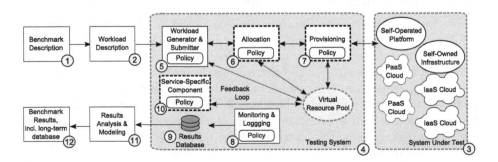

Fig. 1. Overview of our generic architecture for benchmarking IaaS and Paas clouds.

architecture for benchmarking IaaS and PaaS clouds, is depicted in Fig. 1. The main differences between the architecture proposed in this work, and the generic architecture for IaaS benchmarking we have proposed in our previous work, is the focus of the current architecture on *both* IaaS and PaaS cloud elements. The most important conceptual change occurs for component 10, which the current architecture is making aware of the service to be provided, and in particular of the platform configuration and management policies. In Sect. 4, where we adapt this architecture for graph analytics PaaS, we show more practical distinctions.

The *Benchmarking Process* consists of the set of rules, prior knowledge (invariants), and procedures used to subject the SUT to the benchmark workload, and to collect and report the results. In our architecture, the process begins with the user (e.g., a prospective cloud user) defining the benchmark configuration, that is, the complex workloads that define the user's preferred scenario (component 1 in Fig. 1). The scenario may focus on processing as much of the workload as possible during a fixed test period or on processing a fixed-size workload as quickly or cheaply as possible. The benchmarking system converts (component 2) the scenario into a set of workload descriptions, one per (repeated) execution. The workload may be defined before the benchmarking process, or change (in particular, increase) during the benchmarking process. To increase the statistical confidence in obtained results, subjecting the SUT to a workload may be *repeated* or the workload may be *long-running*. The definition of the workload should avoid common pitfalls that could make the workload unrepresentative [20,36].

After the preparation of the workload, the SUT (component 3 in Fig. 1) is subjected to the workload through the job and resource management services provided by the testing system (component 4, which includes components 5–10). In our benchmarking architecture, the SUT can be comprised of one or several self-owned infrastructures, and public and private IaaS clouds; the SUT can also be comprised of self-managed or cloud-provisioned PaaS, which provide services used by the application developed by the customer. For both IaaS and PaaS usage scenarios, the SUT provides resources for the execution of the workload; these resources are grouped into a *Virtual Resource Pool*. The results produced during the operation of the system may be used to provide a *feedback loop* from the Virtual Resource Pool back into the Workload Generator and Submitter (component 5); thus, our architecture can implement open and closed feedback loops [54].

As a last important sequence of process steps, per-experiment results are combined into higher-level aggregates, first aggregates per workload execution (component 11 in Fig. 1), then aggregates per benchmark (component 12). The reporting of metrics should try to avoid the common pitfalls of performance evaluation; see for example [4,22]. For large-scale distributed systems, it is particularly important to report not only the basic statistics, but also some of the outliers, and full distributions or at least the higher percentiles of the distribution (95-th, 99-th, etc.). We also envision the creation of a general database of results collected by the entire community and shared freely. The organization and operation of such a database is within the scope of future work.

3.2 Distinguishing Design Features

We present in the remainder of this section several of the distinguishing features of this architecture.

Commercial clouds may not provide (comprehensive) services for managing the incoming stream of requests (components 5, 6, and 8 in Fig. 1) or the resources leased from the cloud (components 7 and 8). Our architecture supports various policies for provisioning and allocation of resources (components 6 and 7, respectively). Our generic cloud-benchmarking architecture also includes support for evolving black-box systems (components 9, 11, and 12), complex workloads and multi-tenancy scenarios (components 1, 2, and 5), domain-specific components (component 10), etc.

Experiments conducted on large-scale infrastructure should be designed to minimize the time spent effectively using resources. The interplay between components 1, 2, and 5 in Fig. 1 can play a non-trivial role in resolving this challenge, through automatic selection and refinement of complex test workloads that balance the trade-off between accuracy of results and benchmark cost; the main element in a dynamic tuning of this trade-off is the policy present in component 5. The same interplay enables multi-tenancy benchmarks.

Several of the possible SUTs expose complete or partial operational information, acting as white or partially white boxes. Our architecture allows exploiting this information, and combining results from black-box and white-box testing. Moreover, the presence of the increasingly higher-level aggregations (components 11 and 12 in Fig. 1) permits both the long-term evaluation of the system, and the combination of short-term and long-term results. The policy for monitoring and logging in component 8 allows the user to customize which information is collected, processed, and stored in the results database, and may result in significantly lower overhead and, for cloud settings, cost. In this way, our architecture goes far beyond simple black-box testing.

Supports domain-specific benchmarks is twofold in our architecture. First, components 5–7 support complex workloads and feedback loops, and policy-based resource and job management. Second, we include in our architecture a domain-specific component (component 10) that can be useful in supporting cloud programming models such as the compute-intensive workflows and bags-of-tasks, and the data-intensive MapReduce and Pregel. The policy element in component 10 allows this component to play a dynamic, intelligent role in the benchmarking process.

4 Graphalytics: A Graph Analytics Benchmark

In this section, we introduce Graphalytics, a benchmark for graph analytics platforms. Graphalytics is derived from the generic architecture, by adapting and extending several of its key components. Graphalytics includes in the benchmarking process the input dataset, the analytics algorithm, and the provisioned infrastructure. It provides components for platform configuration, deployment, and monitoring, and already provides reference implementations for several popular platforms, such as Giraph and Hadoop/YARN. The Graphalytics API offers a unified execution environment for all platforms, and consistent reporting that facilitates comparisons between all possible combinations of platforms, datasets, and algorithms. The Graphalytics API also permits developers to integrate new graph analytics platforms into the benchmark.

4.1   Design Overview

The design of Graphalytics is based on the generic benchmarking architecture, and on our previous work on benchmarking graph processing systems [25, 27].

The components for benchmark and workload description (components 1 and 2 in Fig. 2, respectively) in the Graphalytics process are derived from our previous work [27] in benchmarking graph-processing platforms. For the benchmark

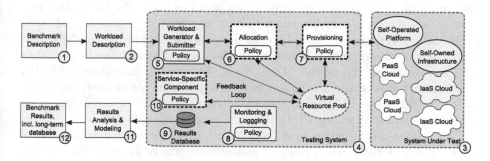

Fig. 2. The Graphalytics architecture for benchmarking graph analytics clouds. The highlighted components indicate the parts of the generic architecture for cloud benchmarking that Graphalytics adapts and extends.

Table 1. Survey of algorithms used in graph analytics. (Source: [27])

Class	Typical algorithms	Number	Percentage [%]
General statistics	Triangulation [60], Diameter [37], BC [56]	24	16.1
Graph traversal	BFS, DFS, Shortest Path Search	69	46.3
Connected components	MIS [9], BiCC [15], Reachability [11]	20	13.4
Community detection	Clustering, Nearest Neighbor Search	8	5.4
Graph evolution	Forest Fire Model [41], Preferential Attachment Model [6]	6	4.0
Other	Sampling, Partitioning	22	14.8
Total		149	100

description (component 1), Graphalytics focuses, process-wise, on processing as many edges or vertices per second, for a variety of input datasets and processing algorithms. The cost of processing, when known, is also of interest for this benchmarking process.

For the workload description (component 2), Graphalytics focuses on algorithms and input datasets. Graphalytics includes already five classes of algorithms, one for each major class of algorithms covered by research published in major conferences by the distributed systems and the databases communities, as indicated by Table 1. More classes of algorithms are currently added to Graphalytics; to keep the number of tested algorithms manageable, we will define more diverse benchmarking scenarios, such that each scenario can focus on a sub-set of existing algorithms.

Component 5 focuses on generating and submitting workloads representative for graph analytics. Graphalytics defines here an archive of commonly used datasets, such as the group considered in our previous work [25]. The archive can grow in time during use, as datasets are either imported, or generated and saved in the database. Currently, Graphalytics includes datasets from the SNAP and GTA [26] collections, and recommends the use of synthetic datasets generated with the Graph500 [23] or the LDBC Social Networking Benchmark [39]. We envision here alternatives to the implementation of this component, especially exploiting the trade-off between fast submission (reads from the database or full-scale generation) and cost (of storage, of computation, etc.).

The service-specific component (component 10) ensures the graph analytics service executes the correct, possibly platform-specific or tuned version of the algorithm, on the input workload provided by the workload generator and submitter (component 5). Additionally, each platform can have specific configuration parameters for the execution of an arbitrary algorithm. Last, the service-specific component checks that the outcome of the execution fulfills the validity criteria of the benchmark, such as output correctness, response time limits, etc.

The monitoring and logging component (component 8) records relevant information about the SUT during the benchmark. Graphalytics supports monitoring and logging for various graph analytics platforms, from the distributed Giraph to the single-node, GPU-based Medusa.

Component 11 enables automated analysis and modeling of benchmarking results. Upon completion of each benchmarking session, the results are aggregated both in human-readable format for direct presentation to the user, and in a standard machine-readable format for storage in the results database. The machine-readable format enables automated analysis of individual platforms, and comparison of results (via component 12). Graphalytics also exposes an API to the platform to enable the platform to signal to the system monitor important events during execution. These events can be superimposed on resource utilization graphs in the benchmark report to enable users to identify bottlenecks and better understand system performance.

Graphalytics includes a centralized repository of benchmark results (component 12), which holds data for various platforms and deployments, and enables comparison of operational data and performance results across many platforms.

The repository offers a comprehensive overview of performance results from the graph analytics community.

4.1.1   Benchmarking with Graphalytics in Practice

We present now the practical steps for benchmarking with Graphalytics.

The benchmark reads the configuration files describing the datasets and platforms to be benchmarked. For datasets, the configuration includes the path to data, the format of the data (e.g., edge-based or vertex-based; directed or undirected; etc.), and parameters to each algorithm (e.g., source vertex for BFS, maximum number of iterations for CD, etc.). Graphalytics includes a comprehensive set of datasets for which configuration files are also provided, requiring no further user input. To run a benchmark on a platform supported by Graphalytics, using a dataset included in Graphalytics, users only need to configure the details of their platform setup, e.g., set the path of HADOOP_HOME for benchmarking Hadoop MapReduce2.

After reading the configuration, Graphalytics runs the platform-specific implementation of each algorithm included in the configuration. For each combination of dataset and algorithm, the graph analytics platform under test is instructed to execute the platform-specific implementation of the algorithm. The core is also responsible for by uploading the selected dataset to the platform.

The system monitor records relevant information about the graph analytics platform under test during the benchmark. After the test, the output validator checks the results to ensure their validity. The report generator gathers for every dataset and algorithm pair the results from the system monitor and the results validator. Users can export these results, and share them with the community through the centralized repository.

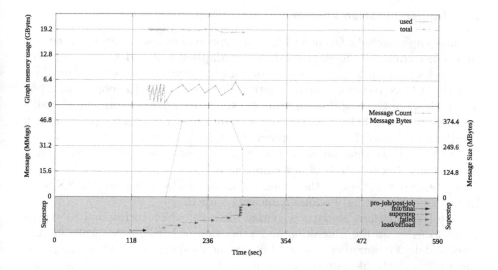

Fig. 3. Resource utilization during a Giraph measurement.

The Graphalytics prototype already includes an implementation of extended monitoring, for the Giraph platform. Figure 3 depicts in its top and middle sub-plots two platform-specific performance metrics collected during a benchmark run, messages sent and memory used, respectively. The bottom sub-plot of the figure depicts the evolution of processing in terms of BSP synchronization super-steps, which characterize the processing model used in Giraph (and Pregel). The supersteps are aligned with the performance metrics and enable the user to asso-ciate the metrics with the evolution of the processing algorithm. For example, the memory usage varies in accordance with the succession of supersteps.

4.1.2   Extensibility of Graphalytics

The aim of the Graphalytics project is to include as many algorithms and plat-forms possible. To accommodate this, the design of the Graphalytics is extensible through simple APIs between components.

Developers can add new graph analytics platforms by providing the following: (a) an implementation of each algorithm defined by the Graphalytics project; (b) functions to upload datasets to the platform, and to download results from the platform; (c) a function to execute a specific algorithm on a specific dataset. Because datasets are managed by the Graphalytics benchmark, no platform-specific changes are needed to support a new dataset; conversely, no new dataset configuration is needed when a new platform is added.

The addition of new algorithms is also supported. Adding a new algorithm requires change in both datasets and platforms. For each dataset, a new set of algorithm-specific parameters needs to be added; a default is provided. For each platform, an implementation of the algorithm must be provided to complete the benchmark.

4.2   Implementation Details

We have implemented a Graphalytics prototype in Java, using .`properties` files for configuration. We include support for MapReduce2 and Giraph, running on Apache Hadoop 2.4.1 or later. Our Giraph implementation of algorithms uses the YARN resource manager introduced in Apache Hadoop 2, avoiding the overhead of the MapReduce framework. We have also experimented, using earlier prototypes of Graphalytics and a simplified benchmarking process, with MapReduce/Hadoop 0.20, Stratosphere 0.2, GraphLab 2.1, and Neo4j 1.5.

In Listing 1, we show an example of a graph configuration file (lines starting with a double back-slash, "\\", are comments). Most elements are optional, e.g., there are no parameters for the connected components algorithm (CONN, lines 19 and 20 in the listing) because none are required.

The benchmarking scenario is `graph.ldbc-30`, which correspond to experi-ments focusing on the generation and validation of LDBC data, scale 30. The generated data is stored as a physical file on the local filesystem where the gener-ator runs, with the file name specified in line 2 of Listing 1. The directedness and the format of the generated dataset are specified in lines 5 and 6, respectively.

```
1   # Filename of graph on local filesystem
2   graph.ldbc-30.file = ldbc-30
3
4   # Properties describing the directedness and format
5   graph.ldbc-30.directed = true
6   graph.ldbc-30.edge-based = true
7
8   # List of algorithms supported on the graph
9   graph.ldbc-30.algorithms = bfs, cd, conn, evo, stats
10
11  # Parameters for BFS
12  graph.ldbc-30.bfs.source-vertex = 12094627913375
13
14  # Parameters for CD
15  graph.ldbc-30.cd.node-preference = 0.1
16  graph.ldbc-30.cd.hop-attenuation = 0.1
17  graph.ldbc-30.cd.max-iterations = 10
18
19  # Parameters for CONN
20
21  # Parameters for EVO
22  graph.ldbc-30.evo.max-id = 2000000000000000
23  graph.ldbc-30.evo.pratio = 0.5
24  graph.ldbc-30.evo.rratio = 0.5
25  graph.ldbc-30.evo.max-iterations = 6
26  graph.ldbc-30.evo.new-vertices = 10
27
28  # Parameters for STATS
29  graph.ldbc-30.stats.collection-node = 12094627913375
```

Listing 1. Sample configuration for an Graphalytics instance.

The benchmarking scenario runs five algorithms, specified on line 9 of
Listing 1. Except for the algorithm for connected components, all other algorithms require one or several parameters. For example, the BFS graph-traversal
algorithm requires a source–the graph vertex where the traversal starts–, which
is specified on line 12.

4.3   Experimental Results

We have used the Graphalytics prototype for two broad types of platforms,
distributed and GPU-based. Exploring the performance of these platforms is
currently a hot topic, with interesting recent results [25,28,42].

We ran Graphalytics using the MapReduce2 and Giraph implementations of
BFS and CONN on the SNB 1000 dataset. We used, for benchmarking these *distributed graph-analytics platforms*, a 10-worker cluster with a separate management node for the HDFS name node, YARN resource manager, and ZooKeeper.
The results are presented in Fig. 4. We have reported much more in-depth results

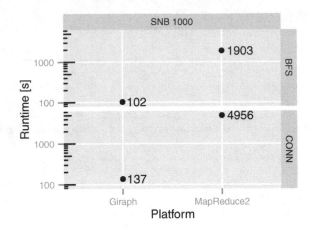

Fig. 4. Algorithm runtimes obtained with Graphalytics.

for other scenarios, but obtained with much more difficulty in setup and execution of the benchmarks because of the lack of tool similar to Graphalytics, in our previous work [25]. We conclude that Graphalytics is a useful benchmark for distributed platforms.

We tested several popular *GPU-based graph analytics platforms* using the LDBC SNB data generator, although the benchmarking code is not yet fully integrated in Graphalytics. Figure 5 depicts the results for the platforms Medusa (M in the figure); Totem (T-G for the Totem version using only the GPU as computing resource, and T-H for the Totem version using both the GPU and the CPU as a hybrid computing resource); and MapGraph (MG). The benchmarking process runs for three GPU-based systems, with GPUs of three different generations and internal architectures. When the integration of this benchmarking process will be complete, Graphalytics will also fully support a variety of graph analytics platforms using GPUs or heterogeneous hardware.

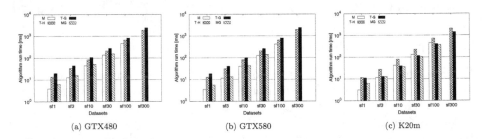

(a) GTX480   (b) GTX580   (c) K20m

Fig. 5. Runtimes of GPU-based platforms executing PageRank on different GPUs and LDBC SNB dataset scales.

5   Benchmarking Challenge: Including the Algorithm in the SUT

Given the extreme scale of the datasets that need to be analyzed, as well as the increasingly complex analysis that needs to be performed, graph analytics has become a high-performance computing (HPC) concern. This trend is probably best proven by the intense activity and fast changes happening in the Graph500[2] ranking, as well as in the adoption of graph traversals as important benchmarks [13] and drivers for irregular algorithms programming paradigms [49].

At the same time, the state-of-the-art in high performance computing is massive parallel processing, backed up by a large variety of parallel platforms ranging from graphical processing units (GPUs) to multi-core CPUs and Xeon Phi. Because traditional graph processing algorithms are known for their parallelism-unfriendly features - data-dependency, irregular processing, bad spatial and temporal locality [1] - a lot of work has focused on developing GPU-specific [10,38,46,52], multi-core CPU-specific [2], or even vendor-specific [14,50] algorithms.

All this work proves an important point: most graph analytics applications can be solved by multiple, different algorithms. These algorithms show very different performance behavior on different platforms *and* on different datasets. Therefore, we argue that not algorithms are the same just because they solve the same problem! Therefore, the selection of the algorithm, which often dictates a pre-selection of the data representation and a filtering of the hardware building blocks, has a huge impact on the observed performance.

To illustrate the importance of the algorithm selection, we present in the remainder of this section a comprehensive evaluation of a large set of 13 parallel breadth-first search (BFS) algorithms built and optimized for GPU execution. A similar study for multi-core CPUs, but not for GPUs, is available in [2]. We point out that the number of algorithms and variations designed for massively parallel architectures such as GPUs is significantly higher, making the results of our GPU study even more interesting.

5.1   Parallel BFS Algorithms for GPU Execution

A BFS traversal explores a graph level by level. Given a graph $G = (V, E)$, with V its collection of vertices and E its collection of edges, and a source vertex s (considered as the only vertex on level 0), BFS systematically explores edges outgoing from vertices at level i and places all their destination vertices on level $i + 1$, *if* these vertices have not been already discovered at prior levels (i.e., the algorithm has to distinguish *discovered* and *undiscovered* vertices to prevent infinite loops).

---

[2] http://www.graph500.org.

5.1.1 Naïve BFS

In a BFS that accepts an edge list as input, an iteration over the entire set of edges is required for each iteration. By a simple check on the source vertex of an edge, the algorithm can determine which edges to traverse, hence which destination vertices to place in the next level of the resulting tree.

Our parallel BFS works by dividing the edge list into sub-lists, which are processed in parallel: each thread will traverse its own sub-list in every iteration. Synchronization between levels is mandatory to insure a full exploration of the current level before starting the next one.

When mapping this parallel kernel to OpenCL, each thread is mapped to an work-item. As global synchronization is necessary, we implement a two-layer barrier structure: first at work-group level (provided by OpenCL), then between work-groups (implemented in-house). This solution limits the synchronization penalty - see [48], Chap. 3 for more details).

Our results - presented below in Sect. 5.2 - show that this approach, although naive, can outperform more optimized versions for specific datasets. Moreover, when comparing it with the naive vertex-centric BFS version included in the Rodinia benchmark [13], the results remain very interesting: in some cases, the vertex-centric version performs better, while in others it is significantly worse than our naive edge-based version [58].

5.1.2 BFS at Maximum Warp

The major performance penalty where running massively parallel traversals on GPUs is work imbalance. This imbalance can appear due to the inability of the programming model to express enough fine-tune options. In [52], the authors present a way to systematically deal with this lack of fine-grain tuning and demonstrate it on a new BFS solution. Specifically, they propose a warp-centric solution, with different parallel stages. This approach enables load-balancing at the finest granularity, thus limiting the penalty of coarser load imbalanced solutions.

We have implemented this approach and varied its granularity. Our results show that, for certain graphs, this algorithm delivers a significant performance boost when compared with the alternatives, but for other datasets it is outperformed by more optimized versions.

5.1.3 The Lonestar Suite

The LonestarGPU collection[3], presented in detail in [10], includes a set of competitive BFS implementations specifically designed to use the processing power of GPUs. The brief descriptions of these algorithms is presented in Table 2.

We point out that each of these algorithms uses different GPU constructs to implement the same BFS traversal. However, depending on the dataset that is being benchmarked, the performance impact of such optimizations varies, showing better results for one or another variant.

---

[3] http://iss.ices.utexas.edu/?p=projects/galois/lonestargpu.

Table 2. Brief description of the LonestarGPU BFS algorithms.

lonestar	A topology-driven, one node-per-thread version.
topology-atomic	A topology-driven version, one node-per-thread version that uses atomics.
merrill	Merrill's algorithm [45].
worklist-w	A data-driven, one node-per-thread version.
worklist-a	A flag-per-node version, one node-per-thread version.
worklist-c	A data-driven, one edge-per-thread version using Merrill's strategy.

5.2   Experiments and Results

We have run 13 different BFS solutions - our edge-based naive version, the maximum warp version with varying warp size (1,2,4,8,16, and 32), and the 6 algorithms from LonestarGPU - on three different GPUs - a GTX480, a C2050, and a K20 m. In this section, we only present our results from the newest GPU architecture in this series, namely the K20 m Kepler-based machine. For the full set of results, we redirect the reader to [44]. The datasets we have selected (from the SNAP repository [40]) are presented in Table 3. We also note that for the `as-skitter` and `roadNet-CA` graphs, we have used both the directed and undirected versions of the graph.

Table 3. The datasets selected for our BFS exploration.

Graph	Vertices	Edges	Diameter	90-percentile Diameter
as-skitter	1,696,415	11,095,298	25	6
facebook-combined	4,039	88,234	8	4.7
roadNet-CA	1,965,206	5,533,214	849	500
web-BerkStan	685,23	7,600,595	514	9.9
wiki-Talk	2,394,385	5,021,410	9	4

Our results on the K20m GPU are presented in Figs. 6 (for the relative performance of the kernels only) and 7 (for the relative performance of the full BFS run, including the data tranfers).

We make the following observations. First, the difference between the best and worst version for each dataset can be as large as *three orders of magnitude*, which means that the choice of algorithm for a given application must be carefully made by the user, prior to measuring and analyzing performance, or be included in the SUT and measured as part of system.

Second, there is no best or worst performing algorithm across all datasets. This indicates that the relative performance of the algorithms also varies with the input data. This variation demonstrates that no single algorithm can be labeled

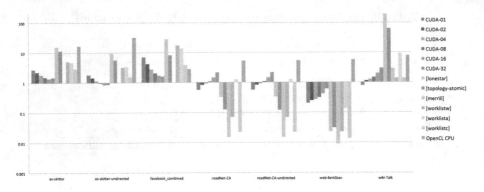

Fig. 6. Parallel BFS performance of the kernel for 13 different GPU implementations. The presented speed-up is normalized to the naive edge-based version (reported as reference, at 1).

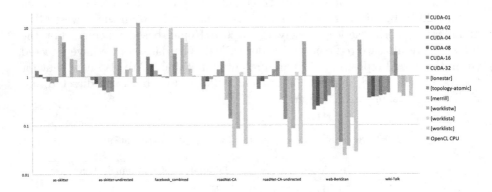

Fig. 7. Parallel BFS performance of the full BFS run (including data transfers to and from the GPU) for 13 different implementations. The presented speed-up is normalized to the naive edge-based version (reported as reference, at 1).

"the best GPU BFS implementation", and indicates that a complete benchmark should include different BFS versions for a comprehensive evaluation.

Third, and final, taking the data transfer times into account levels the differences between different implementations: this special GPU-related overhead is usually so much higher than the execution of the BFS kernel itself, that the small differences between different algorithms become less visible. We argue that this should be reported, for fairness, when comparing against other platforms (see the OpenCL CPU version in Fig. 7), but should be eliminated when benchmarking only GPU-based platforms.

5.3   Practical Guidelines

Our study on BFS running on GPUs demonstrates that the algorithms play an essential role in the performance of the application. Given that the increasing variety of modern parallel architectures leads to an increasing variety of algorithms—some portable, some not—for the same application, the traditional benchmarking approach where the application dictates the algorithm must be deprecated. Much like sorting is a problem with many solutions, many graph applications have many parallel and distributed algorithmic solutions. These algorithms need to be first-class citizens in modern benchmarking.

The main challenge for *algorithm-aware benchmarking* is ensuring completeness. In this context, complete coverage, that is, including all possible algorithms, is not feasible in a reasonable time. Defining the most representative algorithms, which is the typical approach in benchmarking, poses the same problems as defining the most representative applications: it requires criteria that are difficult to define without years of practice and experience with each algorithms. Finally, keeping up-to-date with all the new algorithm developments requires an unprecedented level of implementation and engineering efforts.

Therefore, we believe that an algorithmic-aware benchmarking is necessary, but must be a community-based effort. Without the expertise of different researchers in domains ranging from algorithmics to benchmarking and from single-node to cloud-level platforms, we cannot overcome the technical and completeness challenges that arise in graph processing. Without this effort, the benchmarks we build will be constraint by design.

6   Conclusion and Ongoing Work

The success of cloud computing services has already affected the operation of many, from small and medium businesses to scientific HPC users. Addressing the lack of a generic approach for cloud service benchmarking, in this work we propose a generic architecture for benchmarking IaaS and PaaS clouds. In our generic architecture, resource and job management can be provided by the testing infrastructure, there is support for black-box systems that change rapidly and can evolve over time, tests are conducted with complex workloads, and various multi-tenancy scenarios can be investigated.

We adapt the generic architecture to data-intensive platforms, and design Graphalytics, a benchmark for graph analytics platforms. Graphalytics focuses on the relationship between the input dataset, the analytics algorithm, and the provisioned infrastructure, which it quantifies with diverse performance and system-level metrics. Graphalytics also offers APIs for extending its components, and for supporting more graph analytics platforms than available in our current reference implementation. Experiments we conduct in real-world settings, and with distributed and GPU-based graph analytics platforms, give strong evidence that Graphalytics can provide a unified execution environment for all platforms, and consistent reporting that facilitates comparisons between all possible combinations of platforms, datasets, and algorithms.

Derived from the experience we have accumulated evaluating graph analytics platforms, we identify an important new challenge for benchmarking in clouds: including the algorithm in the system under test, to also benchmark the platform-specific and data-dependent performance of the various algorithms.

We are currently implementing Graphalytics and refining its ability to capture algorithmic bottlenecks. However, to succeed Graphalytics cannot be a single-team effort. We have initiated various community-wide efforts via our work in the SPEC Research Group and its Cloud Working Group, and are currently looking to connect to the SPEC Big Data Working Group.

Acknowledgments. This work is supported by the Dutch STW/NOW Veni personal grants @large (#11881) and Graphitti (#12480), by the EU FP7 project PEDCA, by the Dutch national program COMMIT and its funded project COMMissioner, and by the Dutch KIEM project KIESA. The authors wish to thank Hassan Chafi and the Oracle Research Labs, Peter Boncz and the LDBC project, and Josep Larriba-Pey and Arnau Prat Perez, whose support has made the Graphalytics benchmark possible; and to Tilmann Rabl, for facilitating this material.

References

1. Lumsdaine, B.H.A., Gregor, D., Berry, J.W.: Challenges in parallel graph processing. Parallel Process. Lett. **17**, 5–20 (2007)
2. Agarwal, V., Petrini, F., Pasetto, D., Bader, D.A.: Scalable graph exploration on multicore processors. In: SC, pp. 1–11 (2010)
3. Albayraktaroglu, K., Jaleel, A., Wu, X., Franklin, M., Jacob, B., Tseng, C.-W., Yeung, D.: Biobench: a benchmark suite of bioinformatics applications. In: ISPASS, pp. 2–9. IEEE Computer Society (2005)
4. Amaral, J.N.: How did this get published? pitfalls in experimental evaluation of computing systems. LTES talk (2012). http://webdocs.cs.ualberta.ca/amaral/Amaral-LCTES2012.pptx. Accessed October 2012
5. Amazon Web Services. Case studies. Amazon web site, October 2012. http://aws.amazon.com/solutions/case-studies/. Accessed October 2012
6. Barabási, A.-L., Albert, R.: Emergence of scaling in random networks. Science **286**, 509–12 (1999)
7. Brebner, P., Cecchet, E., Marguerite, J., Tuma, P., Ciuhandu, O., Dufour, B., Eeckhout, L., Frénot, S., Krishna, A.S., Murphy, J., Verbrugge, C.: Middleware benchmarking: approaches, results, experiences. Concurrency Comput. Pract. Experience **17**(15), 1799–1805 (2005)
8. Buble, A., Bulej, L., Tuma, P.: Corba benchmarking: a course with hidden obstacles. In: IPDPS, p. 279 (2003)
9. Buluç, A., Duriakova, E., Fox, A., Gilbert, J.R., Kamil, S., Lugowski, A., Oliker, L., Williams, S.: High-productivity and high-performance analysis of filtered semantic graphs. In: IPDPS (2013)
10. Burtscher, M., Nasre, R., Pingali, K.: A quantitative study of irregular programs on GPUS. In: 2012 IEEE International Symposium on Workload Characterization (IISWC), pp. 141–151. IEEE (2012)
11. Cai, J., Poon, C.K.: Path-hop: efficiently indexing large graphs for reachability queries. In: CIKM (2010)
12. Chapin, S.J., Cirne, W., Feitelson, D.G., Jones, J.P., Leutenegger, S.T., Schwiegelshohn, U., Smith, W., Talby, D.: Benchmarks and standards for the evaluation of parallel job schedulers. In: Feitelson, D.G., Rudolph, L. (eds.) JSSPP 1999, IPPS-WS 1999, and SPDP-WS 1999. LNCS, vol. 1659, pp. 67–90. Springer, Heidelberg (1999)

13. Che, S., Boyer, M., Meng, J., Tarjan, D., Sheaffer, J.W., Lee, S.H., Skadron, K.: Rodinia: a benchmark suite for heterogeneous computing. In: The 2009 IEEE International Symposium on Workload Characterization, IISWC 2009, pp. 44–54 (2009)
14. Checconi, F., Petrini, F.: Massive data analytics: the graph 500 on IBM blue Gene/Q. IBM J. Res. Dev. **57**(1/2), 10 (2013)
15. Cong, G., Makarychev, K.: Optimizing large-scale graph analysis on multithreaded, multicore platforms. In: IPDPS (2012)
16. Deelman, E., Singh, G., Livny, M., Berriman, J.B., Good, J.: The cost of doing science on the cloud: the montage example. In: SC, p. 50. IEEE/ACM (2008)
17. Downey, A.B., Feitelson, D.G.: The elusive goal of workload characterization. SIG-METRICS Perform. Eval. Rev. **26**(4), 14–29 (1999)
18. Eeckhout, L., Nussbaum, S., Smith, J.E., Bosschere, K.D.: Statistical simulation: adding efficiency to the computer designer's toolbox. IEEE Micro **23**(5), 26–38 (2003)
19. Folkerts, E., Alexandrov, A., Sachs, K., Iosup, A., Markl, V., Tosun, C.: Benchmarking in the cloud: what it should, can, and cannot be. In: Nambiar, R., Poess, M. (eds.) TPCTC 2012. LNCS, vol. 7755, pp. 173–188. Springer, Heidelberg (2013)
20. Frachtenberg, E., Feitelson, D.G.: Pitfalls in parallel job scheduling evaluation. In: Feitelson, D.G., Frachtenberg, E., Rudolph, L., Schwiegelshohn, U. (eds.) JSSPP 2005. LNCS, vol. 3834, pp. 257–282. Springer, Heidelberg (2005)
21. Genbrugge, D., Eeckhout, L.: Chip multiprocessor design space exploration through statistical simulation. IEEE Trans. Comput. **58**(12), 1668–1681 (2009)
22. Georges, A., Buytaert, D., Eeckhout, L.: Statistically rigorous java performance evaluation. In: OOPSLA, pp. 57–76 (2007)
23. Graph500 consortium. Graph 500 benchmark specification. Graph500 documentation, September 2011. http://www.graph500.org/specifications
24. Gray, J. (ed.): The Benchmark Handbook for Database and Transasction Systems. Mergan Kaufmann, San Mateo (1993)
25. Guo, Y., Biczak, M., Varbanescu, A.L., Iosup, A., Martella, C., Willke, T.L.: How well do graph-processing platforms perform? an empirical performance evaluation and analysis. In: IPDPS (2014)
26. Guo, Y., Iosup, A.: The game trace archive. In: NETGAMES, pp. 1–6 (2012)
27. Guo, Y., Varbanescu, A.L., Iosup, A., Martella, C., Willke, T.L.: Benchmarking graph-processing platforms: a vision. In: ICPE, pp. 289–292 (2014)
28. Han, M., Daudjee, K., Ammar, K., Özsu, M.T., Wang, X., Jin, T.: An experimental comparison of pregel-like graph processing systems. PVLDB **7**(12), 1047–1058 (2014)
29. Iosup, A.: Iaas cloud benchmarking: approaches, challenges, and experience. In: HotTopiCS, pp. 1–2 (2013)
30. Iosup, A., Epema, D.H.J.: GrenchMark: a framework for analyzing, testing, and comparing grids. In: CCGrid, pp. 313–320 (2006)
31. Iosup, A., Epema, D.H.J., Franke, C., Papaspyrou, A., Schley, L., Song, B., Yahyapour, R.: On grid performance evaluation using synthetic workloads. In: Frachtenberg, E., Schwiegelshohn, U. (eds.) JSSPP 2006. LNCS, vol. 4376, pp. 232–255. Springer, Heidelberg (2007)
32. Iosup, A., Ostermann, S., Yigitbasi, N., Prodan, R., Fahringer, T., Epema, D.H.J.: Performance analysis of cloud computing services for many-tasks scientific computing. IEEE Trans. Par. Dist. Syst. **22**(6), 931–945 (2011)
33. Iosup, A., Prodan, R., Epema, D.: Iaas cloud benchmarking: approaches, challenges, and experience. In: Li, X., Qiu, J. (eds.) Cloud Computing for Data-Intensive Applications. Springer, New York (2015)

34. Iosup, A., Prodan, R., Epema, D.H.J.: Iaas cloud benchmarking: approaches, challenges, and experience. In: SC Companion/MTAGS (2012)
35. Jackson, K.R., Muriki, K., Ramakrishnan, L., Runge, K.J., Thomas, R.C.: Performance and cost analysis of the supernova factory on the amazon aws cloud. Sci. Program. 19(2–3), 107–119 (2011)
36. Jain, R. (ed.): The Art of Computer Systems Performance Analysis. Wiley, New York (1991)
37. Jiang, W., Agrawal, G.: Ex-MATE: data intensive computing with large reduction objects and its application to graph mining. In: CCGRID (2011)
38. Katz, G.J., Kider Jr., J.T.: All-pairs shortest-paths for large graphs on the GPU. In: 23rd ACM SIGGRAPH/EUROGRAPHICS Symposium on Graphics Hardware, pp. 47–55 (2008)
39. LDBC consortium. Social network benchmark: Data generator. LDBC Deliverable 2.2.2, September 2013. http://ldbc.eu/sites/default/files/D2.2.2_final.pdf
40. Leskovec, J.: Stanford Network Analysis Platform (SNAP). Stanford University, California (2006)
41. Leskovec, J., Kleinberg, J.M., Faloutsos, C.: Graphs over time: densification laws, shrinking diameters and possible explanations. In: Proceedings of the Eleventh ACM SIGKDD International Conference on Knowledge Discovery and Data Mining, Chicago, Illinois, USA, pp. 177–187, 21–24 August 2005
42. Lu, Y., Cheng, J., Yan, D., Wu, H.: Large-scale distributed graph computing systems: an experimental evaluation. PVLDB 8(3), 281–292 (2014)
43. Mell, P., Grance, T.: The NIST definition of cloud computing. National Institute of Standards and Technology (NIST) Special Publication 800–145, September 2011. http://csrc.nist.gov/publications/nistpubs/800-145/SP800-145.pdf. Accessed October 2012
44. de Laat, C., Verstraaten, M., Varbanescu, A.L.: State-of-the-art in graph traversals on modern arhictectures. Technical report, University of Amsterdam, August 2014
45. Merrill, D., Garland, M., Grimshaw, A.: Scalable GPU graph traversal. SIGPLAN Not. 47(8), 117–128 (2012)
46. Nasre, R., Burtscher, M., Pingali, K.: Data-driven versus topology-driven irregular computations on GPUs. In: 2013 IEEE 27th International Symposium on Parallel & Distributed Processing (IPDPS), pp. 463–474. IEEE (2013)
47. Oskin, M., Chong, F.T., Farrens, M.K.: Hls: combining statistical and symbolic simulation to guide microprocessor designs. In: ISCA, pp. 71–82 (2000)
48. Penders, A.: Accelerating graph analysis with heterogeneous systems. Master's thesis, PDS, EWI, TUDelft, December 2012
49. Pingali, K., Nguyen, D., Kulkarni, M., Burtscher, M., Hassaan, M.A., Kaleem, R., Lee, T.-H., Lenharth, A., Manevich, R., Méndez-Lojo, M., et al.: The tao of parallelism in algorithms. ACM SIGPLAN Not. 46(6), 12–25 (2011)
50. Que, X., Checconi, F., Petrini, F.: Performance analysis of graph algorithms on P7IH. In: Kunkel, J.M., Ludwig, T., Meuer, H.W. (eds.) ISC 2014. LNCS, vol. 8488, pp. 109–123. Springer, Heidelberg (2014)
51. Raicu, I., Zhang, Z., Wilde, M., Foster, I.T., Beckman, P.H., Iskra, K., Clifford, B.: Toward loosely coupled programming on petascale systems. In: SC, p. 22. ACM (2008)
52. Hong, T.O.S., Kim, S.K., Olukotun, K.: Accelerating CUDA graph algorithms at maximum warp. In: Principles and Practice of Parallel Programming, PPoPP 2011 (2011)
53. Saavedra, R.H., Smith, A.J.: Analysis of benchmark characteristics and benchmark performance prediction. ACM Trans. Comput. Syst. 14(4), 344–384 (1996)

54. Schroeder, B., Wierman, A., Harchol-Balter, M.: Open versus closed: a cautionary tale. In: NSDI (2006)
55. Sharkawi, S., DeSota, D., Panda, R., Indukuru, R., Stevens, S., Taylor, V.E., Wu, X.: Performance projection of HPC applications using SPEC CFP2006 benchmarks. In: IPDPS, pp. 1–12 (2009)
56. Shun, J., Blelloch, G.E.: Ligra: a lightweight graph processing framework for shared memory. In: PPOPP (2013)
57. Spacco, J., Pugh, W.: Rubis revisited: why J2EE benchmarking is hard. Stud. Inform. Univ. 4(1), 25–30 (2005)
58. Varbanescu, A.L., Verstraaten, M., de Laat, C., Penders, A., Iosup, A., Sips, H.: Can portability improve performance? an empirical study of parallel graph analytics. In: ICPE (2015)
59. Villegas, D., Antoniou, A., Sadjadi, S.M., Iosup, A.: An analysis of provisioning and allocation policies for infrastructure-as-a-service clouds. In: 12th IEEE/ACM International Symposium on Cluster, Cloud and Grid Computing, CCGrid 2012, pp. 612–619, Ottawa, Canada, 13–16 May 2012
60. Wang, N., Zhang, J., Tan, K.-L., Tung, A.K.H.: On triangulation-based dense neighborhood graphs discovery. VLDB 4(2), 58–68 (2010)
61. Yigitbasi, N., Iosup, A., Epema, D.H.J., Ostermann, S.: C-meter: a framework for performance analysis of computing clouds. In: 9th IEEE/ACM International Symposium on Cluster Computing and the Grid, CCGrid 2009, Shanghai, China, pp. 472–477, 18–21 May 2009

And All of a Sudden: Main Memory Is Less Expensive Than Disk

Martin Boissier[✉], Carsten Meyer, Matthias Uflacker, and Christian Tinnefeld

Hasso Plattner Institute, Prof.-Dr.-Helmert-Str. 2-3, 14482 Potsdam, Germany
{Martin.Boissier,Carsten.Meyer,Matthias.Uflacker,
Christian.Tinnefeld}@hpi.de

Abstract. Even today, the wisdom for storage still is that storing data in main memory is more expensive than storing on disks. While this is true for the price per byte, the picture looks different for price per bandwidth. However, for data driven applications with high throughput demands, I/O bandwidth can easily become the major bottleneck. Comparing costs for different storage types for a given bandwidth requirement shows that the old wisdom of inexpensive disks and expensive main memory is no longer valid in every case. The higher the bandwidth requirements become, the more cost efficient main memory is. And all of sudden: main memory is less expensive than disk.

In this paper, we show that database workloads for the next generation of enterprise systems have vastly increased bandwidth requirements. These new requirement favor in-memory systems as they are less expensive when operational costs are taken into account. We will discuss mixed enterprise workloads in comparison to traditional transactional workloads and show with a cost evaluation that main memory systems can turn out to incur lower total costs of ownership than their disk-based counterparts.

Keywords: TCO · Mixed workload · Bandwidth · In-memory systems

1 Introduction

Until today the wisdom for storage still is: storing data in main memory is more expensive than storing on disks. Especially with the recent rise of main memory-resident database systems, this cost comparison is often brought up as an argument for disk-based systems.

While this is true comparing the price per byte, the picture looks different if we compare the price per bandwidth. But the price of provided bandwidth from the primary persistence to the CPUs is of high importance as modern applications have increasing bandwidth requirements, e.g., due to demands for big data analytics, real-time event processing, or mixed enterprise workloads including operational reporting on transactional data. When handling such workloads bandwidth easily exceeds single disks. In such cases, local disk arrays or storage servers are employed with large RAID installations in order to meet bandwidth requirements by parallelizing I/O over multiple disks.

© Springer International Publishing Switzerland 2015
T. Rabl et al. (Eds.): WBDB 2014, LNCS 8991, pp. 132–144, 2015.
DOI: 10.1007/978-3-319-20233-4_12

What is missing in our opinion is a thorough analysis that quantifies which costs are to be expected when bandwidth requirements exceed traditional database deployments. With increasingly large setups to provide sufficient I/O bandwidth it is often overlooked how expensive such systems can get, especially when including operational costs.

In this paper, we take a closer look at mixed workloads on relational databases. Mixed workloads combine transactional workloads (OLTP) with analytical workloads (OLAP) on a single system. Basically all major database vendors are currently adding analytical capabilities to their database products to allow operational reporting on transactional data and remove static materialized aggregates [10,12,13], underlining the necessity to research the rather new field of mixed enterprise workloads. Most of these solutions use main memory-resident data structure in order to provide sufficient performance. The reason is that disk-based data structures as they are used in most recent relational database have shown to be too slow compared to main memory-based solutions even when fully cached [7].

Disk-based databases are proven to be a good fit for traditional OLTP workloads, but their viability has to be re-evaluated for mixed workloads. The traditional approach of storing data on disk and caching the most recently used pages in memory is no longer viable as analytical queries often require to process vast numbers of tuples. The requirements of mixed workloads, which are e.g. analytical queries on transactional data – without pre-computed aggregates – increase bandwidth requirements significantly. We think that this trend will continue and steadily increase bandwidth requirements for modern database systems, rendering main memory-based systems increasingly feasible and viable.

Throughout this paper, we will make the following contributions:

- Estimation of the total cost of ownership (TCO) of several system configurations fulfilling certain bandwidth requirements (Sect. 2). We will make a simplified cost calculation to compare HDD-based (hard disk drive-based), SSD-based (solid state disk-based), and MM-based (main memory-based) servers including acquisition costs as well as operational costs.
- Evaluation of bandwidth requirements of a mixed workload using the CH-benCHmark (Sect. 3) for three relational databases.
- A discussion about the break-even point at which main memory-resident databases start to be less expensive than their disk-based counterparts. We project the usage of upcoming enterprise systems and combine it with bandwidth requirements discussed in Sect. 3 to analyze at which point a MM-based system will be the least expensive solution.

We think that this discussion is also important for the field of big data benchmarking. First, even big data veterans as Google recently added transactional features to their F1 database [15], showing that transactional safety might also be important for systems considered as being "big data". Hence, benchmarking enterprise workloads is of increasing relevance. Second, we argue that most benchmarks – TPC-* as well as other current benchmark proposals for big data – need to include operational costs, because our calculations for the three-year costs of large scale-out systems show that operational costs can become the dominating cost driver.

Last but not least, no matter how the application looks, a thorough analysis of bandwidth requirements and the resulting costs are of relevance for any larger server system.

2   Cost Calculations

In order to quantify the costs of bandwidth, we calculate the prices for three different server configurations in this Section: SSD-, HDD-, and MM-based servers. For this calculation, we started with a given bandwidth requirement and a data set size of 500 GB. Depending on the bandwidth we configured the servers using market-available components that are able to fulfil the bandwidth requirements. The HDD- and SSD-based calculations are done using vendor data for maximal bandwidth, not taking into account that these bandwidths are hard to achieve in real-life.

As we are arguing for MM-based systems, we will use more realistic assumptions here in contrast to the other system configurations. Consequently, we do not assume a maximal bandwidth of 80 GB/s per CPU (or 320 GB/s for a four socket node as stated in the technical specifications) but Intel's results for a standardized CPU benchmark.

To be capable of handling actual enterprise workloads, all server configurations have a high-performance PCI-e connected SSD for logging. Obviously, the main memory server also includes sufficient persistent storage to store at least one snapshot of the database.

The main memory size for the SSD- and HDD-based systems is set to ~10 % of the whole data set. In our calculations, the data set size is 500 GB consequently the main memory size is 50 GB.

All server configurations are build using a modern four socket server blade. Since the discussed workloads in this paper are bandwidth bound, the processors for the SSD- and HDD-based servers are Intel Xeon E7-4850v2 CPUs. For the MM-based server we decided for a more expensive CPU with an improved main memory to CPU throughput.

We do not consider possible locking or contention for any of the configurations, neither disk-based systems nor main memory-based systems. We do also not include costs for networking (network cables, switches, et cetera).

High-Availability and Durability. For all configurations and calculations we include both the costs for acquisition as well as operational costs. Furthermore, we include costs for high availability. For each configuration, we assume a system to be highly available when one server node can fail as there is an additional spare node in each configuration. I.e., for a single server node configuration high availability can increase total costs by a factor of two.

Durability issues and failures of single components as hard disks or DRAM chips are not considered.

2.1   Server Configurations

Main Memory-Based System. The assumed main memory-based server is a four socket system equipped with four Intel Xeon E7-4890v2 CPUs. Recent benchmarks by Intel have shown that such a system achieves a bandwidth of up to ~246 GB/s for the STREAM Triad Benchmark[1]. Any bandwidth exceeding the maximum of 246 GB/s requires a multi-node setup to scale.

We assume that the memory size has to be at least a factor of two larger than the data set. This space is required for the operation system (OS), intermediate results, et cetera. Consequently, the main memory-based systems includes 1 TB of DRAM for the 500 GB data set, a PCIe-connected SSD for logging, and a 500 GB HDD for database persistence (e.g., snapshotting).

HDD-Based System. The HDD-based server is a four socket node equipped with four Intel Xeon E7-4850v2 CPUs. The size of main memory is set according to Oracle's MySQL sizing documents [16], which recommend to reserve main memory to cache 5 %-10 % of the data set size. Consequently, we assume 25 GB main memory for database and 25 GB for the operation system (i.e., 50 GB).

The disks are put in direct-attached storage (DAS) units, where each DAS unit contains up to 96 disks in a RAID array. Two SAN controllers are used to connect each DAS unit. This setup yields a bandwidth of 6 GB/s (each SAN controller has a theoretical bandwidth of 3 GB/s) per DAS unit. Per server node up to eight SAN controllers can be used. Consequently, the peak bandwidth per HDD-based server is 24 GB/s ($4 * 2 * 3\,GB/s$).

The systems adapts to increasing bandwidth requirements first by adding DAS units and then by adding server nodes.

SDD-Based System. The solid state disk-based systems are configured using SSDs that are connected via the PCI Express (PCIe) bus as the latest generation of SSD exceeds the bandwidth of SATA ports. Each SSD provides a read bandwidth of 3 GB/s. Using Intel Xeon E7-4850v2 CPUs we assume that each socket can directly connect two PCIe 16 × SSDs at full bandwidth. For a four socket server a maximum of eight PCIe-connected SSDs can be used, yielding a peak bandwidth of 24 GB/s.

The systems adapts to increasing bandwidth requirements first by adding PCIe SSDs and then by adding server nodes.

2.2   TCO Calculations

We calculated the TCO for the three server configurations using varying bandwidth requirements.

For all configurations we assume a data set size of 500 GB. Even with low bandwidth requirements this configuration is already comparatively expensive

---

[1] Intel Xeon E7-4890v2 Benchmark – URL: http://www.intel.com/content/www/us/en/benchmarks/server/xeon-e7-v2/xeon-e7-v2-4s-stream.html.

in a main memory-based configuration, because the server has to be able to store the whole data set in main memory.

Please note, that the sum for the high availability costs shown in Fig. 1 represent the costs for the added server node including operational costs. The other two sums (acquisitional and operational costs) show the total costs for the system without high availability.

Configuration 1 - 10 GB/s Bandwidth. In case of a required bandwidth of 10 GB/s a HDD-based system is the less expensive solution (see Fig. 1(a)). The SSD-based server has significantly lower operational costs, but suffers from high availability costs as well as the MM-based solution. The MM-based server is comparatively expensive since – as stated above – main memory has to be sufficiently large to store the data set entirely in memory. In our calculations that results in a memory size of 1 TB.

For a fair comparison, it has to be said that the server configurations are built to scale to higher bandwidth requirements. Especially for a bandwidth requirement as 10 GB/s there are more price efficient configurations for disk-based setups.

Configuration 2 - 20 GB/s Bandwidth. The picture looks different if we assume a bandwidth of 20 GB/s (see Fig. 1(b)). The HDD-based solution is still the least expensive one, but the main memory-based solution is already less expensive than the SSD-based server.

The main cost driver for SSD-based over MM-based systems are expensive PCIe-connected flash drives of which seven are required to theoretically provide the required bandwidth. Even though PCIe-connected SSDs outperform HDDs, the performance to price ratio is not significantly improved compared to recent 15 K HDDs.

Configuration 3 - 40 GB/s Bandwidth. As shown in Fig. 1(c), for a bandwidth requirement of 40 GB/s, the costs for a MM-based solution are lower than for the other two configurations. With this bandwidth requirement, costs for HDD-based servers are clearly dominated by the operational costs while SSD-based server are again dominated by the costs for PCIe-connected flash drives.

3   Bandwidth Requirements for Enterprise Workloads

To better understand the TCO for a given bandwidth requirement we measured the reading bandwidth of mixed enterprise workloads. Mixed workloads include transactional workloads as they are executed daily in business systems and further include analytical workloads, which are usually handled by data warehouses. Mixed workload systems are currently being researched intensively (e.g., HYRISE [5], H-Store [6], HyPer [9]) and are also the focus of several commercial products (e.g., SAP HANA [4], Oracle Times Ten, IBM DB2 BLU [13]).

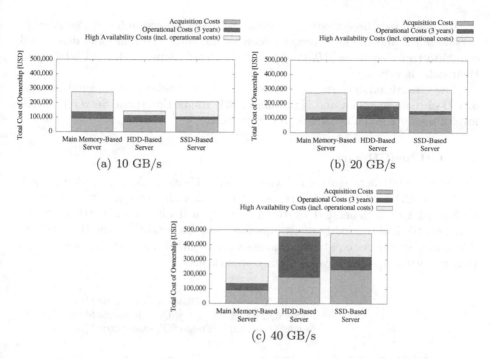

Fig. 1. Total costs of ownership for the server configurations (see Sect. 2.1) with varying bandwidth requirements.

They all aim at introducing analytical real-time capabilities into transactional systems.

To quantify how much read bandwidth is required executeing mixed workloads we chose the CH-benCHmark [2] as the assumed workload. The CH-benCHmark is a modification of the TPC-C benchmark. To include analytical characteristics of mixed workloads the CH-benCHmark runs analytical queries adopted from the TPC-H benchmark on the transactional TPC-C data set.

3.1 Setup

We created a TPC-C data set with a scale factor of 1000. The data set was generated using the OLTP-Bench framework [3] and has a size of ~70–85 GB (depending on the database used). We used *iotop*[2] to record reads from disk. Execution times for the analytical CH-benCHmark queries are not discussed in this paper as our focus is on the required bandwidth. We did not consider further characteristics as latency or IOPS (input/output instructions per second) due to our focus on mixed workloads, which are read-dominated. Also, we expect main memory-based systems to outperform disk-based systems for these characteristics anyhow.

---

[2] iotop – URL: http://guichaz.free.fr/iotop/.

We evaluated three relational databases: MySQL version 5.6.4, PostgreSQL version 9.3.4, and the most recent release of a commercial relational disk-based database (referred to as *DBMS X*). For all installations, we left all settings to their default values.

The virtualized database server has one terabyte network connected storage, 32 GB of main memory, and is running SUSE Linux Enterprise Server 11 patch level 2 respectively Windows Server 2008 RC2.

3.2   CH-benCHmark

Because the CH-benCHmark is based on TPC-C, we decided to measure each analytical CH-benCHmark query on its own in order to quantify how much additional I/O is generated by running analytical queries on a transactional data set. We had to modify several queries to be executable on the three chosen databases. Whenever non-trivial changes were required (see Sect. 7.1) we skipped that particular query to avoid unfair comparisons.

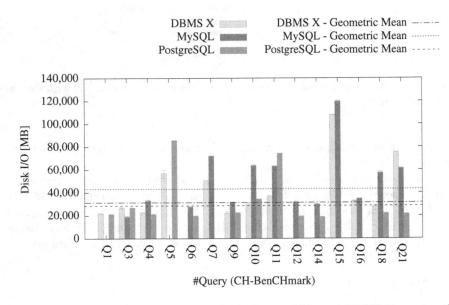

Fig. 2. Megabytes read from disk for MySQL, PostgreSQL, and DBMS X running the analytical queries of the CH-benCHmark. Geometric means are calculated using all queries that ran on all three databases (i.e., Q3, Q4, Q9, Q10, Q11, Q18, and Q21).

We executed each query five times with one non-included execution upfront to warm up the database and OS file caches. Figure 2 shows the measured average data transfers of each query execution with error bars showing the minimum/maximum measurement of the five executions. For each database, a vertical line shows the geometric means of read data for all queries that ran on all three databases (i.e., Q3, Q4, Q9, Q10, Q11, Q18, and Q21) (Table 1).

Table 1. Geometric mean and average of megabytes read from disk for MySQL, PostgreSQL, and DBMS X running the analytical queries of the CH-benCHmark.

Database	Data Read from Disk (MB)	
	Geometric Mean	Average
DBMS X	31,368.7	34,373.7
MySQL	43,258.9	47,055.2
PostgreSQL	28,550.3	31,832.4

4   Discussion

The next generation of enterprise systems will have vastly different characteristics compared to today's systems. In 2014, Plattner presented such an (already productive) enterprise system [12]. These new systems have to handle an increasing analytical pressure without materialized aggregates in order to provide higher flexibility and fewer restrictions on analytical queries, thus executing all calculations directly on the transactional data. Today, basically all major database vendors are working on techniques to provide sufficient performance for mixed workloads [4,10,13]. Hence, we think that mixed workload benchmarks as the CH-benCHmark are an important indicator of upcoming workload requirements and are well worth examining.

To discuss the break-even point of bandwidth requirements and main memory-based servers, let us assume an enterprise system with 500 parallel users and a data set size of 1 TB. Each user executes one analytical query on the database system every two hours, resulting in 250 analytical queries per hour. Assuming the average query reads of 31 GB (see Sect. 3.2), this scenario would have a bandwidth requirement of ~21.5 GB/s. As shown in Sect. 2, we see that SSD-based servers are less expensive in this scenario. However, neither the number of 500 parallel users and especially not the database size of 1 TB is what counts as a "large enterprise system" today. Especially the assumption of one analytical query per user every two hours is rather conservative. If we assume 500 parallel users executing analytical queries once an hour, the bandwidth requirements would already favor main memory-based solutions.

5   Related Work

To our best knowledge, there is no previous work that has yet discussed cost effectiveness of main memory databases for high-bandwidth workloads. However, there are several publications that discuss the link between architectural decisions and their impacts on costs.

Rowstron et al. presented a similar argument to ours for big data analytics [14]. The authors argue that scaled-up servers with large main memory volumes are often a better solution than disk-based Hadoop clusters. This observation is especially interesting as Rowstron et al. point out that the median MapReduce job size of Yahoo and Microsoft is smaller 14 GB and that 90 % of Facebook's jobs process less than 100 GB of data.

Another related topic is how to improve operational costs for server systems. While this topic is well researched for traditional architectures, it is rather new to systems where massive main memory sizes are responsible for a large share of the energy consumption. Malladi et. al discussed this topic and found that energy can be decreased by a factor 3–5 without sacrificing too much performance [11]. However, lowering the power of main memory also lowers the maximal bandwidth.

Another very interesting aspect – one that is not covered in this paper – was discussed by Zilio et al. [17]. They argue that modern software systems have become increasingly sophisticated requiring several domain experts to handle and tune the systems: "These economic trends have driven the total cost of ownership of systems today to be dominated by the cost of people, not hardware or software". Of course, such a metric is hard to quantify. But it can also be seen as arguing in our favor because an often proposed advantage of main memory systems is the simplification of system architectures [1].

6   Future Work

There is still a considerable amount of work ahead of us. The main topics that we want to continue working on are:

Bandwidth Evaluation. To further evaluate bandwidth requirements we want to examine the CH-benCHmark from end to end, including bandwidth requirements for transactional queries as well as for analytical queries.

Besides the already evaluated databases, we are planning to benchmark other databases to gain a better overview. These alternatives include IBM DB2 with its columnar accelerator BLU [13] and MonetDB [8], both representing disk-based open sourced databases.

We expect columnar databases as MonetDB to require less I/O when executing analytical queries. However, it will be interesting how high the I/O overhead for tuple reconstruction using columnar data structures (and compression) is. Furthermore, it is interesting to see how the usage of compression effects read bandwidth.

General Workload Assumptions. It is very hard to estimate and predict how workloads might look if databases are capable of mixed workloads and high performance analytics. We want to talk to experts and find out which workload assumptions are realistic for the next years and how workloads might look in the future.

Query Cost Model. In the long run, we want to work towards a general cost model for bandwidth requirements of workloads. Even though we think that a holistic view over several different databases is already very helpful, there are still many variables in each implementation that are hard to factor out.

Emerging Memory / Storage Technologies. The performance developments of disk-based storage and main memory will probably lower the break-even point even more as the gap between both is currently still widening. New developments as non-volatile memory (NVM) thrive to increase the density of byte-addressable storage significantly, potentially having huge impacts on the TCO of database systems with high bandwidth requirements.

7 Conclusion

The bandwidth evaluations of mixed enterprise workloads in this paper have shown that the requirements of upcoming enterprise systems might very well have completely different bandwidth requirements compared to current enterprise workloads. Comparing disk- and main memory-resident databases in respect to bandwidth shows that main memory-resident databases are not as expensive as often expected. In fact, main memory can be the least expensive storage medium. We think it is import to convey a new point of view in which main memory-based solutions are not "the most expensive solution only viable when everything else is too slow" but rather "the least expensive solutions when performance requirements are high".

Appendix

7.1 Execution of CH-benCHmark Queries

The following adaptions have been done to run the CH-benCHmark queries:

- when needed, the extract function (e.g., `EXTRACT(YEAR FROM o_entry_d)`) has been replaced by the year function (e.g., `YEAR(o_entry_d)`)
- for MySQL and PostgreSQL, query 15 has been modified to use a view instead of using SQL's `having` clause (code provided by the OLTP-Bench framework)
- when needed, aliases have been resolved in case they are not supported in aggregations

We set the maximal query execution time to 12 h for each query, which excludes queries from our results even though they are executable. Due to their long execution time we assume that the execution of these queries does not terminate.

7.2 TCO Calculations

The following section lists the components for an assumed bandwidth requirement of 40 GB/s. The prices have been obtained from the official websites of hardware vendors and do not include any discounts. Energy costs are calculated using the technical specifications of the hardware. Cooling costs are calculated using an assumed Power Usage Effectiveness (PUE) of 1.8 according to the Uptime Institute 2012 Data Center Survey[3]. The cost of energy is $0,276 per kWh. Both energy and cooling costs are calculated for a timespan of three years.

---

[3] Uptime Institute 2012 Data Center Survey – URL: http://uptimeinstitute.com/ 2012-survey-results.

Table 2. TCO calculation for the HDD-based system.

Item	Amount	Est. price per item ($)	Total ($)
Server	2	30,000	60,000
DAS unit	7	4,500	31,500
SAS controller	14	500	7,000
Hard disk (10 K)	672	150	100,800
Main memory (16 GB)	4	340	1,360
Energy	-	-	143,110
Cooling	-	-	114,488
SSD for logging	1	5,000	5,000
TCO	**463,218**		

For the hard disk and solid state disk based systems each node is a four processor server (4 × Intel Xeon E7-4850v2 12C/24T 2.3 GHz 24 MB) with an estimated price of $30,000. For both configurations the size of main memory is set to ∼10 % of the database volume (i.e., 50 GB for the 500 GB data set).

All following exemplary calculations do not include costs for high availability.

HDD-Based System. The HDD-based system adapts to higher bandwidth requirements by adding direct attached storage units. In this calculation, each node has eight SAS slots. Each DAS unit is connected to two SAS slots and is assumed to provide the maximal theoretical throughput of 6 GB/s and consists of 96 disks (10 K enterprise grade) to provide the bandwidth. It is possible to reach 6 GB/s with fewer 15 K disks, but a configuration with 10 K is more price efficient.

Since two SAS slots are used to connect each DAS unit, each server node can connect to a maximum of four DAS units resulting in a peak bandwidth of 24 GB/s. Consequently, any bandwidth higher than 24 GB/s requires an additional server node.

The hardware setup for the 40 GB/s configuration and its TCO calculation is listed in Sect. 7.2 (Table 2).

SSD-Based System. The SSD-based system uses PCI-e connected solid state disks. Recent Intel Xeon CPUs have up to 32 PCI-e lanes per socket that are directly connected. Consequently, we assume a theoretical setup of up to eight PCIe-connected SSDs per server node.

For our calculations, we use an PCIe SSD that provide a peak read bandwidth of 3 GB/s and has a size of 1 TB. As of now, there are faster SSDs available (up to 6 GB/s), but these are more expensive by a factor of over 3x. We also calculated prices for another PCIe SSD vendor whose drives are almost a factor 2x less expensive in their smallest size of 350 GB. We did not include these calculations here, as these drives are currently not available.. However, even using these drives the 40 GB/s configuration is still more expensive than its main memory-based counterpart (Table 3).

Table 3. TCO calculation for the SSD-based system.

Item	Amount	Est. price per item ($)	Total ($)
Server	2	30,000	60,000
PCIe-connected SSD (3 GB/s)	14	13,100	183,400
Main memory (16 GB)	4	340	1,360
Energy	-	-	34,240
Cooling	-	-	27,218
SSD for logging	1	5,000	5,000
TCO	**311,218**		

Main Memory-Based System. The main memory-based server is equipped with Intel's latest XEON E7 CPU. A server with four CPUs (Intel Xeon E7-4890v2 15C/30T 2.8 GHz 37 MB) costs ~$63,000. The costs include a 600 GB enterprise-grade HDD for persistence (Table 4).

Table 4. TCO calculation for the main memory-based system.

Item	Amount	Est. price per item ($)	Total ($)
Server	1	63,000	63,000
Hard disk (15 K)	1	300	150
Main memory (16 GB)	63	340	21,420
Energy	-	-	25,304
Cooling	-	-	20,242
SSD for logging	-	-	5,000
TCO	**135,116**		

References

1. Boissier, M., Krueger, J., Wust, J., Plattner, H.: An integrated data management for enterprise systems. In: ICEIS 2014 - Proceedings of the 16th International Conference on Enterprise Information Systems, vol. 3, 27–30 April, pp. 410–418, Lisbon, Portugal (2014)
2. Cole, R., Funke, F., Giakoumakis, L., Guy, W., Kemper, A., Krompass, S., Kuno, H.A., Nambiar, R.O., Neumann, T., Poess, M., Sattler, K.-U., Seibold, M., Simon, E., Waas, F.: The mixed workload ch-benchmark. In: DBTest, p. 8. ACM (2011)
3. Difallah, D.E., Pavlo, A., Curino, C., Cudr-Mauroux, P.: OLTP-bench: an extensible testbed for benchmarking relational databases. PVLDB **7**(4), 277–288 (2013)
4. Färber, F., May, N., Lehner, W., Große, P., Müller, I., Rauhe, H., Dees, J.: The SAP HANA database - an architecture overview. IEEE Data Eng. Bull. **35**(1), 28–33 (2012)

5. Grund, M., Krueger, J., Plattner, H., Zeier, A., Cudr-Mauroux, P., Madden, S.: HYRISE - a main memory hybrid storage engine. PVLDB **4**(2), 105–116 (2010)
6. H-Store Documentation: MapReduce Transactions. http://hstore.cs.brown.edu/documentation/deployment/mapreduce/
7. Harizopoulos, S., Abadi, D.J., Madden, S., Stonebraker, M.: OLTP through the looking glass, and what we found there. In: SIGMOD Conference, pp. 981–992. ACM (2008)
8. Idreos, S., Groffen, F., Nes, N., Manegold, S., Mullender, K.S., Kersten, M.L.: MonetDB: two decades of research in column-oriented database architectures. IEEE Data Eng. Bull. **35**(1), 40–45 (2012)
9. Kemper, A., Neumann, T., Finis, J., Funke, F., Leis, V., Muehe, H., Muehlbauer, T., Roediger, W.: Processing in the hybrid OLTP & OLAP main-memory database system hyper. IEEE Data Eng. Bull. **36**(2), 41–47 (2013)
10. Larson, P., Clinciu, C., Fraser, C., Hanson, E.N., Mokhtar, M., Nowakiewicz, M., Papadimos, V., Price, S.L., Rangarajan, S., Rusanu, R., Saubhasik, M.: Enhancements to SQL server column stores. In: Proceedings of the ACM SIGMOD International Conference on Management of Data, SIGMOD 2013, June 22–27, pp. 1159–1168, New York (2013)
11. Malladi, K.T., Lee, B.C., Nothaft, F.A., Kozyrakis, C., Periyathambi, K., Horowitz, M.: Towards energy-proportional datacenter memory with mobile dram. In: SIGARCH Computer Architecture News, vol. 40(3), pp. 37–48 (2012)
12. Plattner, H.: The impact of columnar in-memory databases on enterprise systems. PVLDB **7**(13), 1722–1729 (2014)
13. Raman, V., Attaluri, G.K., Barber, R., Chainani, N., Kalmuk, D., KulandaiSamy, V., Leenstra, J., Lightstone, S., Liu, S., Lohman, G.M., Malkemus, T., Müller, R., Pandis, I., Schiefer, B., Sharpe, D., Sidle, R., Storm, A.J., Zhang, L.: DB2 with BLU acceleration: so much more than just a column store. PVLDB **6**(11), 1080–1091 (2013)
14. Rowstron, A., Narayanan, D., Donnelly, A., O'Shea, G., Douglas, A.: Nobody ever got fired for using hadoop on a cluster. In: Proceedings of the 1st International Workshop on Hot Topics in Cloud Data Processing, HotCDP 2012, pp. 2:1–2:5. ACM, New York (2012)
15. Shute, J., Vingralek, R., Samwel, B., Handy, B., Whipkey, C., Rollins, E., Oancea, M., Littlefield, K., Menestrina, D., Ellner, S., Cieslewicz, J., Rae, I., Stancescu, T., Apte, H.: F1: a distributed SQL database that scales. PVLDB **6**(11), 1068–1079 (2013)
16. Sizing Guide for Single Click Configurations of Oracles MySQL on Sun Fire x86 Servers. www.oracle.com/technetwork/server-storage/sun-x86/documentation/o11-133-single-click-sizing-mysql-521534.pdf
17. Zilio, D.C., Rao, J., Lightstone, S., Lohman, G.M., Storm, A.J., Garcia-Arellano, C., Fadden, S.: DB2 design advisor: integrated automatic physical database design. In: VLDB, pp. 1087–1097. Morgan Kaufmann (2004)

FoodBroker - Generating Synthetic Datasets for Graph-Based Business Analytics

André Petermann[1,2](✉), Martin Junghanns[1], Robert Müller[2], and Erhard Rahm[1]

[1] University of Leipzig, Leipzig, Germany
{petermann,junghanns,rahm}@informatik.uni-leipzig.de
[2] Leipzig University of Applied Sciences, Leipzig, Germany
mueller@fbm.htwk-leipzig.de

Abstract. We present FoodBroker, a new data generator for benchmarking graph-based business intelligence systems and approaches. It covers two realistic business processes and their involved master and transactional data objects. The interactions are correlated in controlled ways to enable non-uniform distributions for data and relationships. For benchmarking data integration, the generated data is stored in two interrelated databases. The dataset can be arbitrarily scaled and allows comprehensive graph- and pattern-based analysis.

Keywords: Synthetic data generation · Benchmarks · Graph data management · Business intelligence

1 Introduction

The operations of a company are reflected in its business processes and their different domain objects such as employees, products or purchase orders. Significant goals for business intelligence are to find correlations between such domain objects, to identify how domain objects are involved in business processes and to determine how this affects the success or failure of processes and the company. For example, consider trading processes where unfriendly sales people will have a negative influence while fast logistics companies can have a positive impact. Business information systems supporting the execution of business processes store those domain objects influencing the process outcome as master data. Further on, they record transactional data such as sales orders or invoices referencing master data during process execution.

While the data of business information systems is usually stored in relational databases, all process-related domain objects and their references can be abstracted as a graph, where objects are nodes and references are edges. So far, graph data models have been typically used to model natural graph scenarios such as social networks or knowledge models (e.g., ontologies). However, recent research projects aim at using graph models for data integration and analytics, for example in the enterprise [10,13] or health care [9] domains.

© Springer International Publishing Switzerland 2015
T. Rabl et al. (Eds.): WBDB 2014, LNCS 8991, pp. 145–155, 2015.
DOI: 10.1007/978-3-319-20233-4_13

Evaluating or benchmarking such analytical graph applications requires datasets that reflect the nature of business information systems. Unfortunately, it is very difficult to obtain real datasets from enterprises for evaluation purposes so that we see a need to generate appropriate datasets synthetically. Such datasets must meet specific features to be useful for graph-based analytics which are not sufficiently supported by established data generators for warehouse and graph benchmarks. In particular, the scenario covered by the dataset should represent a large number of similar use cases in practice and should allow the analysis of complex relationships and patterns to gain significant business insights. More specifically, we pose the following requirements.

1 - Heterogeneous Domain Objects. Domain objects belong to different classes representing master or transactional data. Example master data classes are employee, customer or product and example transactional classes are sales order, rating or email.

2 - Heterogeneous Relationships. Domain objects may be related in different semantic ways represented by different relationship types. Relationships have to involve both master and transactional domain objects in any combination.

3 - Support of Graph-Based Analysis. The recorded data should allow a more comprehensive, graph-based business analysis than with traditional data warehouses. For enterprise data, it should thus represent different business processes with complex, correlated interactions between master data objects and transactional objects. It should thus be possible to identify and analyze different transactional patterns, e.g., the effect of how an employee interacted with specific customers.

4 - Multiple Sources. Companies and organizations typically use multiple information systems. To cover data integration tasks, a synthetic dataset needs to imitate multiple data sources.

5 - Scalability. For benchmarking, the datasets need to be scalable up to realistic sizes with thousands of master data objects and millions of transactions.

To our knowledge, there is no data generator that provides all of the stated requirements yet. The contribution of this paper is FoodBroker, a data generator for related master and transactional data which are meaningful for a specific domain. FoodBroker is based on simulating processes and process-supporting information systems. The current version of FoodBroker provides an authentic data model representing two interrelated business information systems and generates records by simulating many business process executions (*cases*). FoodBroker considers the correlation between master data instances and their influence on the development and outcome of each case. The resulting data can be integrated using a graph model and used for evaluation and benchmarking of analytical graph systems. The source code of the FoodBroker data generator can be found on GitHub[1] under GPL v3.

The remaining paper is organized as follows. In Sect. 2 we discuss related work about existing benchmark-related data generators. Then in Sect. 3, we introduce

---

[1] https://github.com/dbs-leipzig/foodbroker.

our business process and its simulation. The current implementation is described in Sect. 4. Finally, we end up with a summary and provide an outlook in Sect. 5.

2   Related Work

Data generators for established OLAP benchmarks such as TPC-H [12] and APB-1 [8] do not fully meet the introduced requirements for graph-based analytics. Although they are scalable and offer heterogenous domain objects their relationship heterogenity is limited. Graph patterns of interest usually involve causal connections among transactional data as well as the involved master data [10]. However, APB-1 and all TPC benchmarks except TPC-DS are focussed on a single class of transactional data and TPC-DS by design provides no relationships in between the different transactional objects. The data generator of BigBench [5] extends the TPC-DS data model by related transactional data from web logs and reviews. However, those relationships are generated without considering the impact of master data instances. Further on, none of the benchmarks involves the problem of integrating multiple sources.

Due to the growing research interest in graph database systems, many benchmark studies have been published comparing those systems to relational database systems as well as among themselves. As there is currently no standard graph benchmark available, varying data generators have been developed, typically focused on a specific use case.

In [14], Vicknair et al. compare a relational and a graph database system. Their generated datasets contain artificial provenance information modeled as a directed acyclic graph and stored in a single data source. Nodes as well as edges are homogeneous and lack specific semantics: nodes carry random payload data, edges have no relationship type and no data attached to them. Graph sizes can vary depending on a user-defined number of nodes.

Holzschuher and Peinl also compare a relational and a graph database system focusing on the evaluation of graph query languages and SQL [7]. The generated datasets resemble the structure and content of online social networks in a property graph and are stored in a single data source. Nodes and edges can be of different types, nodes store realistic data based on dictionaries. The relationship information is taken from a real social network which makes it impossible to scale the graph size unrestricted.

Dominguez-Sal et al. [4] benchmark different GDBMS. They make use of the recursive matrix (R-MAT [3]) algorithm, which was designed to generate directed graphs that are equivalent to real networks in terms of specific graph invariants, like degree distribution and diameter. Using a single parameter, the graph size can be scaled exponentially. The algorithm is focused on generating a realistic structure but lacks the capability to add semantics to nodes or edges.

In [6], Gupta emphasizes that the characteristics of generating data for heterogenous graphs strongly differs from data generators of benchmarks for data warehouses and graphs which are specified by topology characteristics. He proposes a data generator for heterogenous multigraphs with typed nodes and labeled edges representing meaingful data from a drug discovery scenario. Although this approach even provides correlations between domain objects and graph structure, the resulting data does neither consider the characteristics of process-related data nor the scenario of multiple data sources.

Pham et al. propose S3G2, a framework for specifying the generation of graphs using a rule-based approach which leads to plausible structural correlations between graph structures and domain objects [11]. Their focus is the generation of social networks with real-world structural and semantic characteristics using dictionaries and manually defined correlation rules. To achieve scalability in terms of graph size and computing time, the framework is implemented using the MapReduce paradigm. A way to adopt the framework for business analytics could be the definition of domain specific correlation rules. Nevertheless, this would not lead to the generation of complex business processes where the correlation is not only depending on directly connected nodes but on transitive relationships between multiple domain objects.

A promising project, which adopts the S3G2 framework, is the Linked Data Benchmark Council (LDBC) [1,2], which aims at establishing standardized graph-oriented benchmarks and data generators for different application domains. Like us, the authors also highlight the importance of semantic correlations within the data to address business intelligence and graph analytics. Up until now, the project offers data generators for social network and semantic publishing use cases. FoodBroker describes a complementary, business-oriented use case where data objects are interrelated within diverse business process executions. The Foodbroker dataset will thus likely allow different graph-based analysis tasks for business intelligence than in the LDBC use cases. Still we expect the LDBC results as a valuable input for defining a benchmark based on Foodbroker datasets.

3   Simulation

The FoodBroker simulation reflects the core business of a fictive company trading food between producers (vendors) and retailers (customers). The company only offers a brokerage service and has no warehouse. Process-related data is recorded in an enterprise resource planning (ERP) system and a customer issue tracking (CIT) system. The simulation includes the customizable generation of master data as well as transactional data involved in the executions of two interrelated business processes for food brokerage and complaint handling. In the following, we describe the data model and the two simulated business processes. We also discuss how the generated data can be used for graph-based business analytics.

3.1   Data Model

The data model for the ERP and CIT systems is shown in Fig. 1. The ERP system stores master data objects of the classes Employee, Product, Customer, Vendor and Logistics. Products are categozied into different product categories, such as

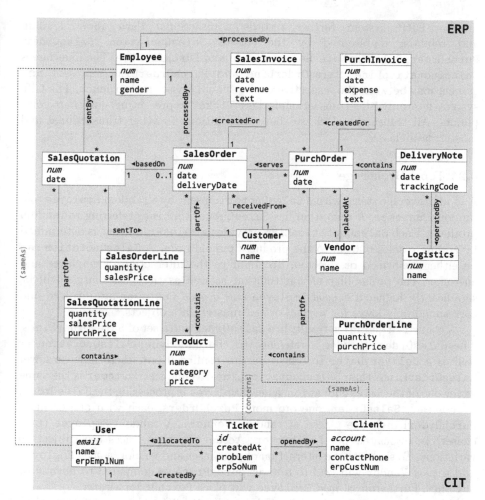

Fig. 1. FoodBroker Data Model: The outer rectangles show the boundaries of two systems ERP and CIT. Database tables correspond either to classes or n:m associations (*Line). Primary keys are highlighted by italic letters. Associations are shown as solid lines. Foreign keys are attached to associations. Implicit associations in between both databases are represented by dotted lines. For each implicit association, there is a corresponding column with prefix erp.

fruits, vegetables and *nuts*. In the CIT system, the instances of master class User refer to employees in the ERP system, while master class Client refers to ERP class Customer.

For each master data object we provide a quality criterion with one of the values *good, normal* or *bad*. We use these criteria in the generation of business processes and achieve thus a correlation between the different kinds of master objects and the structure and outcome of process executions.

The ERP system further records transactional data about trading and refunds, represented by the classes SalesQuotation, SalesOrder, PurchOrder, DeliveryNote, SalesInvoice and PurchInvoice. The line classes SalesQuotationLine, SalesOrderLine and PurchOrderLine represent n:m associations between the respective transactional classes and Product. The CIT system has only the transactional class Ticket representing customer complaints. All transactional classes have associations to other transactional and master data classes.

3.2 Food Brokerage

A food brokerage starts with a SalesQuotation sent by a random Employee to a random Customer. A quotation has SalesQuotationLines referring to random products. Each SalesQuotationLine provides a salesPrice that is determined by adding a sales margin to the products purchPrice. A SalesQuotation can be either confirmed or rejected. To simulate the interaction of employee and customer, the probability of confirmation as well as the sales margin will be significantly higher if a *good* Employee and a *good* Customer are involved and correspondingly lower for *normal* or *bad* master data objects.

A confirmed quotation results in a SalesOrder and a set of SalesOrderLines. The SalesOrder includes a reference to the underlying SalesQuotation as well as a deliveryDate. To reflect partial confirmations, there may be fewer SalesOrderLines than SalesQuotationLines. While the Customer is the same as the one of the SalesQuotation, a new Employee is processing the SalesOrder.

For each SalesOrder, one ore more PurchOrders, each with one or more PurchOrderLines, are placed at random Vendors. A random Employee (purchaser) is associated per PurchOrder. Actual purchase prices are subject to variations. To simulate the interaction of purchaser and vendor, a *good* Employee and a *good* Vendor will lead to a lower purchPrice as compared to *normal* or *bad* ones. Furtheron, a *good* Employee will place PurchOrders faster.

After a PurchOrder is processed by the Vendor, the company receives information about the DeliveryNote, in particular date of delivery, operating Logistics company and operator-specific trackingCode. The delivery time is influenced by the quality of both Vendor and Logistics company such that *good* business partners will lead to faster delivery than *bad* or *normal* ones.

Finally, one SalesInvoice per SalesOrder will be sent to the Customer and one PurchInvoice per PurchOrder received from the corresponding Vendor. All transactional data objects created within cases of food brokerage refer to their predecessor (except SalesQuotation) and provide a date property with a value greater or equal than the one of the predecessor.

3.3 Complaint Handling

For every customer complaint an instance of Ticket referring the corresponding SalesOrder is created. For the first complaint per Customer, additionally a Client instance is created. The employee handling the complaint is recorded in class User. There are two problems which may cause complaints: late delivery and bad product quality. Late delivery complaints occur if the date of a DeliveryNote is greater than the agreed deliveryDate of the SalesOrder, e.g., due to a *bad* Employee processing the SalesOrder or a PurchaseOrder, a *bad* Vendor, a *bad* Logistics company or combinations of those. Bad quality complaints may be caused by *bad* Products , a *bad* Vendor, a *bad* Logistics company or combinations of such.

A Ticket may lead to refunds which are recorded as SalesInvoice objects with negative revenue. While the constellation of a *good* Employee allocated to the Ticket and a *good* Customer may lead to low or no refund, *bad* ones lead to higher refund. If the problem was caused by the Vendor, there will be also a refund for the company in the form of a PurchInvoice with negative expenses. While the constellation of a *good* Employee and a *good* Vendor may lead to high refund, *bad* ones lead to lower or no refund.

3.4 Graph Analytics

The FoodBroker datasets provide complex correlations between master data instances and their involvement within the execution of the described business processes making it a good basis for graph-based business analytics. In particular it is possible to analyze the influence of the different master objects (employees, customers,vendors, etc.) on the financial outcome of business processes by case-wise aggregating all revenue- and expense-related properties, e.g. in SalesInvoice and PurchInvoice records of the same case, as well as determining the impact of refunds due to complaints.

Since all objects involved in the same case are interconnected in the generated dataset, it is possible to analyze the resulting graphs representing business process executions. Hence it is not only possible to aggregate profit-related measures but also to analyze the transactional correlation patterns underlying positive or negative cases to find frequent patterns in such business process executions. For example, it could be found that cases with outstandingly high profit frequently contain patterns such as Customer:ACME <-receivedFrom- SalesOrder -processedBy-> Employee:Alice saying that employee Alice was processing the sales order by customer ACME.

Finding such correlations is a non-trivial task due to many-to-many associations between cases and master data objects of the same class. For example, in a single case multiple Employees can be involved in different ways but also in the same way, e.g. multiple purchaser employees for different items.

4   Implementation

Our current implementation stores the generated dataset in a MySQL database dump with separate databases for the ERP and CIT systems. In both databases every class has a dedicated table with the same name. 1:1 and 1:m associations

Table 1. FoodBroker Configuration Parameters

Class	Configuration	Default Value
Master data		
Employee	number of instances	$30 + 10 \times SF$
	proportions of good/normal/bad inst.	0.1/0.8/0.1
Product	number of instances	$1000 + 10 \times SF$
	proportions of good/normal/bad inst.	0.1/0.8/0.1
	list price range	0.5..8.5
Customer	number of instances	$50 + 20 \times SF$
	proportions of good/normal/bad inst.	0.1/0.8/0.1
Vendor	number of instances	$10 + 5 \times SF$
	proportions of good/normal/bad inst.	0.1/0.8/0.1
Logistics	number of instances	$10 + 0 \times SF$
	proportions of good/normal/bad inst.	0.1/0.8/0.1
Food brokerage		
Process	number of cases	$10000 \times SF$
	date range	2014-01-01..2014-12-31
SalesQuotation	products per quotation (lines)	1..20
	quantity per product	1..100
	sales margin/IM	0.05/0.02
	confirmation probability/IM	0.6/0.2
	line confirmation probability	0.9
	confirmation delay/IM	0..20/5
SalesOrder	agreed delivery delay/IM	2..4/1
	purchase delay/IM	0..2/2
	invoice delay/IM	0..3/−2
PurchOrder	price variation/IM	0.01/−0.02
	delivery delay/IM	0..1/1
	invoice delay/IM	2..5/3
Complaint handling		
Ticket	probability of bad quality/IM	0.05/−0.1
	sales refund/IM	0.1/−0.05
	purchase refund/IM	0.1/0.05

SF abbreviates scale factor
number of instance configurations are defined by linear functions $a + b \times SF$
IM abbreviates impact per master data instance
$IM > 0$ means good/bad master data instances increase/decrease value
$IM < 0$ means good/bad master data instances decrease/increase value

correspond to foreign keys in class tables where the column names represent the relationship type. M:n associations are stored in separate tables.

The simulation is implemented as a Java console application. The database dump includes SQL statements for the creation of tables. All instances of a class are inserted into the tables corresponding to their class. Domain-specific string values such as employee or product names are provided by an embedded SQLite database.

Scalability in terms of different data sizes is controlled by a scale factor (SF). This makes it possible to create datasets with equal criteria regarding distributions, value ranges and probabilities but with a different number of instances. The number of master data instances and the number of simulated cases are defined by linear functions. The simulation function has a default slope of 10,000 per scale factor. We take into account that master data instances do not scale proportionally to cases. For example, there is only a limited amount of logistics companies. Thus, the functions of all master data classes beside a specific slope also have a y-intercept to specify a minimum number of instances.

Beside the scale factor and the resulting data growth, the generation of master data but also the process simulation is customizable by several configuration parameters. For example, one can set the confirmation probability of quotations or the influence of master data quality criteria on that probability. All configurations are set within a single file. An overview about the configuration parameters is provided in Table 1. While using FoodBroker for benchmarks requires fixed configurations and variable scale, variable configurations at a fixed scale can be used to evaluate business intelligence applications at different scenarios. Table 2 shows dataset measures for different scale factors using the standard configuration. With respect to runtime, our current implementation shows a linear behavior for increasing scale factors. All datasets were generated on a workstation containing an Intel Xeon Quadcore, 8 GB RAM and a standard HDD.

Table 2. Measures of FoodBroker datasets for different scale factors (SF)

| SF | Master data | | Transactional data | | | |
	Objects	Time	Objects	Relationships	Time	Dump Size
1	1.1 K	4 s	73 K	380 K	4 s	42 MB
10	1.7 K	4 s	725 K	3.8 M	25 s	426 MB
100	6.8 K	4 s	7.2 M	38 M	4 min	4.1 GB
1000	67 K	8 s	68 M	360 M	35 min	39 GB

5   Summary

We presented the FoodBroker data generator for graph-based analytics based on simulated business process executions. FoodBroker creates domain-specific master and transactional data as well as meaningful relationships in between them.

Quality criteria provided for all master data objects influence the execution of business processes and thus the creation of transactional data and process outcome. The complex correlations between master and transactional objects within graph-like business process representations make the Foodbroker datasets a good candidate for a variety of graph- and pattern-based analysis tasks.

Data generation is customizable and can be scaled to large sizes by a simple scale factor. In future work, we want to define a benchmarking workload for Foodbroker datasets and use it to evaluate different systems including current GDBMS and our own BIIIG^2 approach [10].

Acknowledgments. This work is partly funded within the EU program *Europa fördert Sachsen* of the European Social Fund.

References

1. Angles, R., et al.: The linked data benchmark council: a graph and RDF industry benchmarking effort. ACM SIGMOD Rec. **43**(1), 27–31 (2014)
2. Boncz, P.: LDBC: benchmarks for graph and RDF data management. In: Proceedings of the 17th International Database Engineering and Applications Symposium. ACM (2013)
3. Chakrabarti, D., Zhan, Y., Faloutsos, C.: R-mat: a recursive model for graph mining. In: SDM, vol. 4, pp. 442–446. SIAM (2004)
4. Dominguez-Sal, D., Urbón-Bayes, P., Giménez-Vañó, A., Gómez-Villamor, S., Martínez-Bazán, N., Larriba-Pey, J.L.: Survey of graph database performance on the HPC scalable graph analysis benchmark. In: Shen, H.T., Pei, J., Özsu, M.T., Zou, L., Lu, J., Ling, T.-W., Yu, G., Zhuang, Y., Shao, J. (eds.) WAIM 2010. LNCS, vol. 6185, pp. 37–48. Springer, Heidelberg (2010)
5. Ghazal, A., et al.: Bigbench: towards an industry standard benchmark for big data analytics. In: Proceedings of the 2013 international conference on Management of data. ACM
6. Gupta, A.: Generating large-scale heterogeneous graphs for benchmarking. In: Rabl, T., Poess, M., Baru, C., Jacobsen, H.-A. (eds.) WBDB 2012. LNCS, vol. 8163, pp. 113–128. Springer, Heidelberg (2014)
7. Holzschuher, F., Peinl, R.: Performance of graph query languages: comparison of cypher, gremlin and native access in Neo4j. In: Proceedings of the Joint EDBT/ICDT 2013 Workshops. ACM (2013)
8. OLAP Council.: APB-1 OLAP Benchmark. http://www.olapcouncil.org/research/bmarkly.htm
9. Park, Y., et al.: Graph databases for large-scale healthcare systems: a framework for efficient data management and data services. In: IEEE 30th International Conference on Data Engineering Workshops (ICDEW) (2014)
10. Petermann, A., Junghanns, M., Müller, R., Rahm, E.: BIIIG : enbabling business intelligence with integrated instance graphs. In: IEEE 30th International Conference on Data Engineering Workshops (ICDEW) (2014)
11. Pham, M.-D., Boncz, P., Erling, O.: S3G2: a scalable structure-correlated social graph generator. In: Nambiar, R., Poess, M. (eds.) TPCTC 2012. LNCS, vol. 7755, pp. 156–172. Springer, Heidelberg (2013)

---

^{2} http://www.biiig.org.

12. Transaction Processing Performance Council.: TPC Benchmarks. http://www.tpc. org/information/benchmarks.asp
13. Vasilyeva, E., et al.: Leveraging flexible data management with graph databases. In: 1st International Workshop on Graph Data Management Experiences and Systems. ACM (2013)
14. Vicknair, C., et al.: A comparison of a graph database and a relational database: a data provenance perspective. In: Proceedings of the 48th annual Southeast regional conference. ACM (2010)

Author Index

Printed in the United States
By Bookmasters